North American Mixing Forum

AA003365

Core Programming Topic at the 2012 AIChE Annual Meeting

Pittsburgh, Pennsylvania, USA
28 October - 2 November 2012

ISBN: 978-1-62276-735-9

Printed from e-media with permission by:

Curran Associates, Inc.
57 Morehouse Lane
Red Hook, NY 12571

Some format issues inherent in the e-media version may also appear in this print version.

Printed by Curran Associates, Inc. (2013)

AIChE
3 Park Avenue
New York, NY 10016-5991

Phone: (203) 702-7660
Fax: (203) 775-5177

www.aiche.org

Additional copies of this publication are available from:

Curran Associates, Inc.
57 Morehouse Lane
Red Hook, NY 12571 USA
Phone: 845-758-0400
Fax: 845-758-2634
Email: curran@proceedings.com
Web: www.proceedings.com

TABLE OF CONTENTS

(40a) Transitional Flow Behaviour in the Shear and Sedimentation Test Cell (SSTC)

Marcio Bezerra Machado, Chemical and Materials Engineering, University of Alberta, Edmonton, AB, Canada and Suzanne Kresta, University of Alberta, Edmonton, AB, Canada

The dispersion and dissolution of chemical additives into liquids needs to be efficient in order to achieve reasonable industrial yields. It is known that current bench scale tests do not always represent industrial scale behavior, and turbulence inhomogeneity in bench scale stirred tanks can lead to significant uncertainties about the process mixing requirements. In order to better represent the industrial scale, a new bench scale stirred tank has been developed: the Shear and Sedimentation Test Cell (SSTC). The SSTC provides more uniform intensity of mixing over the volume of the tank, uses a much smaller volume of sample than the traditional stirred tank, and allows for the possibility of a sedimentation step after the mixing step with no sample transfer required. The unique geometry of the SSTC allows a more uniform turbulence distribution, which is not obtained in the other bench scale tests. The SSTC is a narrow and tall tank (T=7.6 cm, H=3T) running with multiple impellers. In order to check the limit of fully turbulent flow in the SSTC, three sets of impellers were used: the A310, Rushton and Intermig. Sets of five (Rushton and A310) or six (Intermig) impellers of a relatively large diameter (D=T/2 for Rushton and A310 and D=2T/3 for Intermig) were used. Two fluids were used in the tank: water (1cp) and triethylene glycol (6 cp); to check the effect of the fluid viscosity. The Reynolds number was varied from 1100 to 30000. The power consumption of all sets of impellers was determined over a range of Reynolds numbers. The dissolution of solids and blending of liquids performance was also assessed. Mean and rms velocities were measured throughout the tank using an LDV. When Rushton impellers were used, the impeller discharge stream was still operating in turbulent conditions at Re = 1 800. The A310 and Intermig were still operating in turbulent flow at Re = 4200-7500. As the impeller to tank diameter ratio was large, the ratio between the maximum local dissipation and the bulk dissipation was smaller than the ratio obtained in a regular mixing tank. The maximum local energy dissipation was located in the impeller discharge stream for all configurations. The SSTC is able to provide a more uniform mixing for the bitumen extraction operation and is proposed as a test cell for additive chemicals in a wide range of applications.

AICHE Annual Conference
Pittsburgh, Oct. 28 – Nov. 2, 2012

NEW IMPELLER DESIGN: ANTI-RAGGING IMPELLER, ARI2

Robert Higbee[a], Jason Giacomelli[a], Wojciech Wyczalkowski[a]

[a] *Philadelphia Mixing Solutions, 1221 East Main st. Palmyra PA 17078, USA;*

jgiacomelli@philamixers.com

bhigbee@philamixers.com

wwyczalkowski@philamixers.com

Abstract.
Environmental mixing applications exist, such as in waste water treatment facilities, where the fluid can have a high fibrous rag-like content. Mixing impellers that resist the accumulation of rags are known as "anti-ragging". Such applications also require the mixing impeller to deliver the maximum amount of axial flow per unit cost of energy. When considering large impellers, for example 2 meters in diameter, fabricated impellers can offer increased design flexibility resulting in lower production costs as opposed to fibreglass reinforced plastic which is used in the current Philadelphia Mixing Solutions Ltd anti-ragging impeller design. The challenge was to design a large diameter, high efficiency, anti-ragging, fabricated blade. The design effort has produced a highly skewed, forward raked, high pumping capacity, low power, fabricated impeller with applications that extend beyond the waste water treatment industry. The impeller design discussion includes the methodology used to allow the trailing edge of the non-helical formed blade to conform to the angles found on a pure helical geometric surface of similar diameter. The characteristic dimensionless impeller constants power and flow number, N_P and N_Q, were measured in the laboratory and compared to other standard impeller types in terms of Energetic and Power efficiency as well as by the cost of flow. N_P was measured over the turbulent flow regime using a rotationally free-floating motor affixed to a load cell at a known radius. N_Q was measured in the turbulent regime using 2D Particle Image Velocimetry (PIV) which was validated by using a flow balance to check the PIV data.

Keywords: Water Treatment, ragging, high efficiency, impeller, PIV

1. IMPELLER DESIGN

This presentation pertains to a specific impeller – The Philadelphia Mixing Solutions ARI2 (Figure 1). The overarching theme is that given a well defined impeller design envelope, physical mathematics can be used to design a specific impeller mechanical shape, to be confirmed by testing, that meets not only process and mechanical strength goals, but also meets fabrication cost goals.

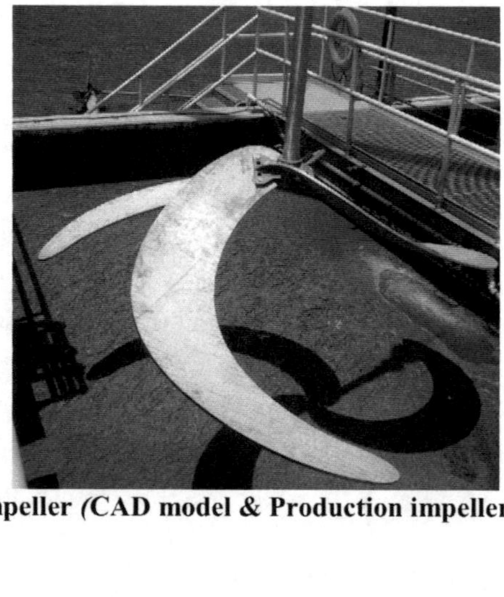

Figure 1. PMSL ARI2 high efficiency anti-ragging impeller (CAD model & Production impeller)

The specific process category being considered is waste water treatment and specifically, that part of the waste water treatment process that contains a significant quantity of stringy, fibrous, rag-like material which can accumulate on impellers that are not specifically designed for this type of service.

Mixing impellers that resist the accumulation of rags are known as "anti-ragging". A high efficiency anti-ragging impeller exists - the Philadelphia Mixing Solutions Ltd (PMSL) fibreglass reinforced plastic (FRP) XELSC-2200 (Figure 2). Its commercial production requires a computer numerically controlled (CNC) machined production mould. This allows the impeller pitch face to have complex cambered geometry. However, the design must be based on the maximum torque in a pre-defined operating range, therefore rendering the impeller less cost effective for the minimum torque application, also the cost of CNC machined moulds increase geometrically as impeller diameter increases.

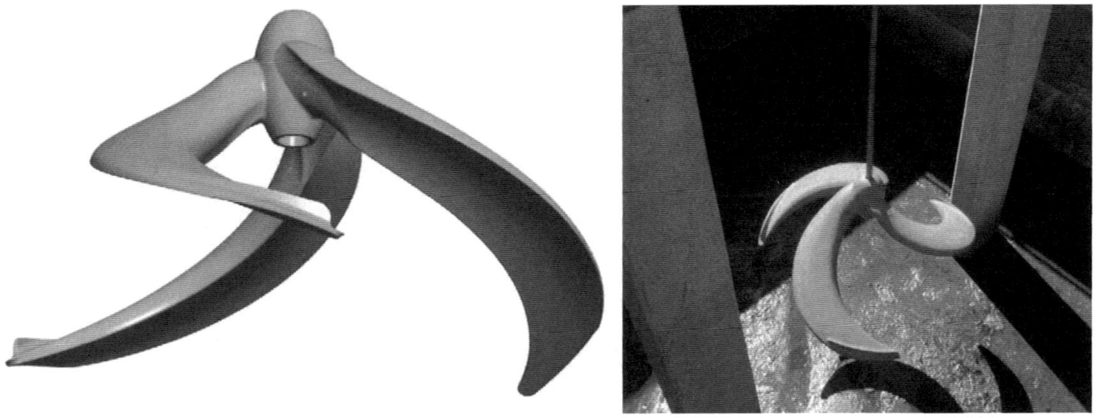

Figure 2. PMSL XELSC-2200 high efficiency anti-ragging impeller developed in 2008. (CAD model & Production impeller)

Fabricated metal impellers are made from flat stock which is available in a variety of standard thicknesses and can be readily formed using standard commercially available "brakes". This allows the fabricated impeller to be optimally designed for the specific required torque which can, in many instances, reduce cost. The design goal of the ARI2 impeller was to utilize metal fabrication technology to create an impeller with a level of efficiency and anti-ragging properties that were equal to or greater than the existing PMSL XELSC-2200 impeller.

Non-proprietary aspects of the ARI-2 design include the concepts of hand, skew, rake, camber, and pitch. The proprietary aspects of this design are best described in terms of the fabrication steps:

1. Cut a uniquely devised profile from flat sheet-metal stock.
2. Apply a cylindrical roll of a specific magnitude (R) at a unique angle (A1).
3. Attached (3) formed blades to a standard 3 X 39 deg PMSL impeller hub at a unique angle (A2) with respect to the hub ear mounting face.

To determine the values of R, A1, & A2; a mathematical optimization study was formulated which considered a design matrix of (9) hypothetical impellers where for each of (3) roll angle parameter magnitudes, the pitch variance optimization goal parameter was computed at (3) different roll radii. The purpose of the optimization study was to find those values of R, A1, and A2 that allow the angles at the blade trailing edge, whose fundamental blade shape is not helical in nature, to conform to the angles found on a pure helical surface of similar diameter and at the desired helical pitch. The pitch variance shown in Figure 3 indicates the conformance of the design to the desired impeller pitch - the lower the pitch variance, the higher the conformance.

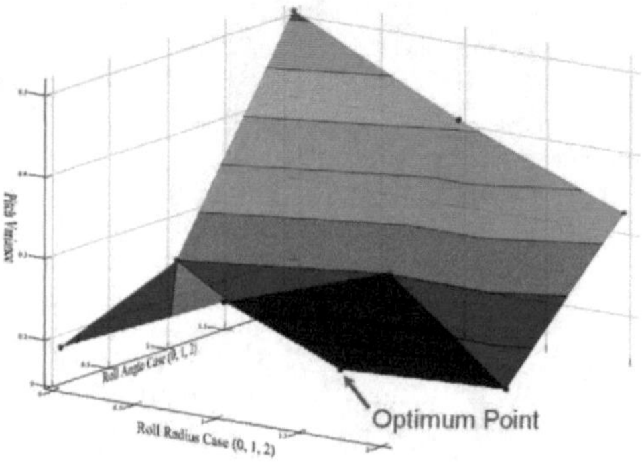

Figure 3. Optimization study results to determine R, A1, and A2 combination that best conforms to helical pitch.

2. PERFORMANCE EVALUATION

2.1 Characteristic Impeller Constants

The conceptual and theoretical design process is concluded by producing several impeller design instances which are converted into physical artifacts via a rapid prototyping process called *Selective Laser Sintering* (SLS). This technique allows the inexpensive production of scaled down prototypes which can be tested in 0.3 – 1m diameter vessels (Figure 4).

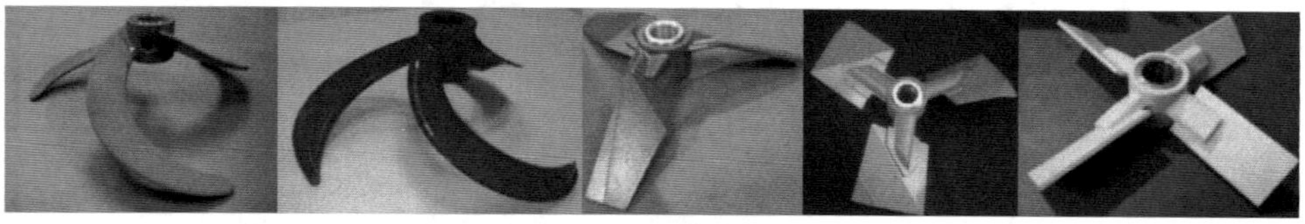

Figure 4. Rapid prototypes of impellers used in lab-scale testing: ARI2; Comparison Impellers: XELSC-2200, low and high solidity hydrofoils, and 4 X 45 deg PBT

The impeller power number, N_P, was measured in a fully baffled vessel over the turbulent flow regime using a rotationally free-floating motor affixed to a load cell at a known radius. Torque was obtained from the product of load cell force and the known radius. A tachometer was used to measure impeller speed. The ARI-2 anti-ragging impeller was measured at several speeds corresponding to turbulent Reynolds numbers. The impeller flow number, N_Q, is measured using a standard 2-D *Particle Image Velocimetry* (PIV) system. A 203mm diameter prototype ARI2 was used for the power and flow measurements. A flow balance around the impeller region was calculated to validate the PIV data. The flow and power number of the ARI2 impeller is presented below in Figure 5 and compared to the common PMSL impellers shown in Figure 4.

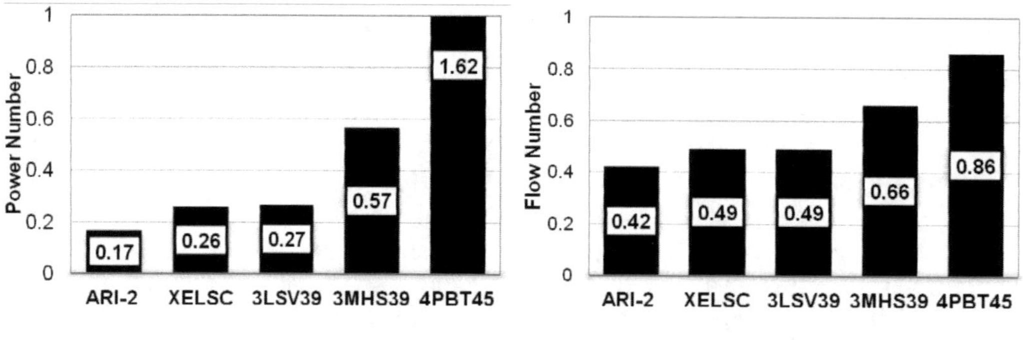

Figure 5. Power and Flow numbers of ARI2 compared to other common impellers.

2.2 Impeller Performance Evaluation

The goal of the impeller design process was to produce an impeller that was less costly to make for a wider range of applications and have comparable efficiency to the XELSC-2200. To evaluate the new impeller against the XELSC-2200, a criterion for comparison was established based on the process requirements. The primary target application of the ARI2 is for anoxic digesters and waste water treatment (WWTR) basins where gentle mixing is required at low energy cost. In environmental applications such as these, flow generated by the impeller and power consumed are the important factors of comparison. Referring to *Hydraulic Efficiency* proposed by Fort [1] and *Power Efficiency* [2] proposed by the BHR Group Limited[2], two equations are available which compare impeller flow number to power number which can be used to evaluate the ARI2 (Equations (1, 2)). The resulting efficiency calculations are shown in Figure 6.

$$Hydraulic\ Efficiency = \frac{N_Q^3}{N_p} \quad (1)$$

$$Power\ Efficiency = \frac{P_{OUT}}{P_{IN}} = \frac{8}{\pi^2}\frac{N_Q^3}{N_p} \quad (2)$$

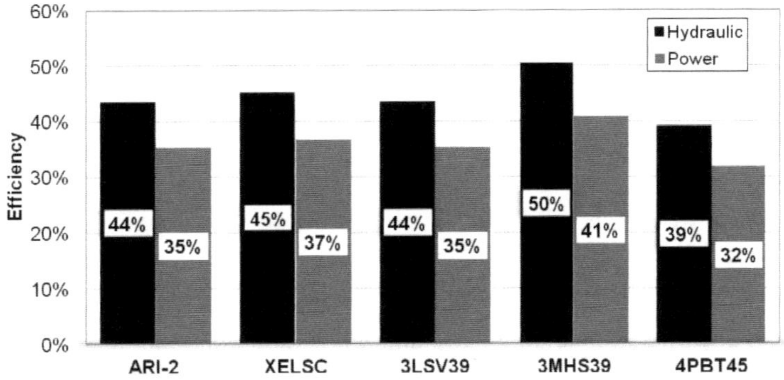

Figure 6. Comparison between Hydraulic Efficiency [1] and Power Efficiency [2].

The ARI2 efficiencies are equal to that of the XELSC-2200 and in the same ranges for similar hydrofoils and turbines. A more process specific criterion to evaluate the ARI2 impeller design is to compare flow generated to the cost, which in this case is directly related to power. By using the expressions for impeller power draw and impeller flow generated, an expression is developed to quantify this cost (Equation (3)).

$$\frac{Flow}{Power} = \frac{Q}{P} = \frac{N_Q N D^3}{N_p \rho N^3 D^5} = \frac{N_Q}{N_p}\left[\frac{1}{\rho N^2 D^2}\right] \quad (3)$$

Considering the unknown variables in these equations, a basis was assumed. An actual application was referenced from the Philadelphia Mixing Solutions environmental team with a 2.54m impeller diameter and 30 RPM (0.5RPS) impeller speed. The flow generated and the power required for each of the impellers can be calculated (Figure 7-Basis 1). Assuming the same diameter but instead varying the impeller speed so that each impeller produces equal flow, a second measure of comparison is obtained (Figure 7-Basis 2).

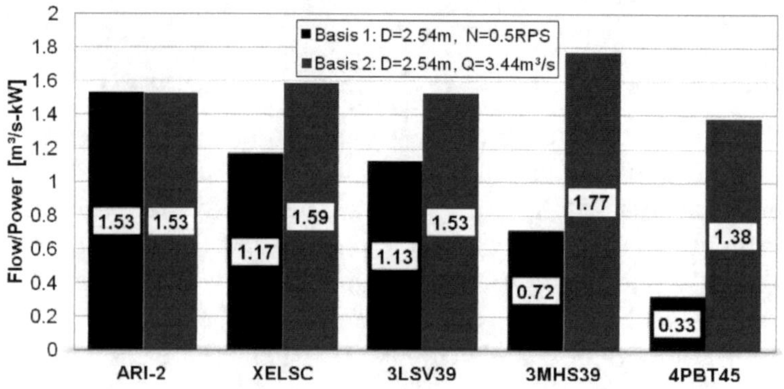

Figure 7. Flow per unit cost calculated by holding impeller diameter constant by either varying impeller flow for equal impeller speed (Basis 1) or varying impeller speed to achieve equal flow (Basis 2).

The conclusion drawn from Figure 7 is that if the ARI2 is to be retrofitted to an existing motor/gearbox in a WWTR basin, an equal diameter impeller will result in a higher flow and mixing and will consume no extra power. In contrast, if a new installation is the case, a smaller piece of equipment can be used which will reduce operating and capital costs but will not sacrifice mixing and process performance.

3. APPLICATION AND FUNCTION

The second purpose of this impeller design is to resist the buildup of fibrous materials found in waste water treatment basins known as *rags*. These materials accumulate on the impeller blades over time and gradually increase the impeller power draw until the motor trips. The process vessel must then be drained and the rags removed from the impeller causing undesired downtime and maintenance costs.

A lab simulation was used to measure ragging performance using scaled down parameters from a full scale WWTR application based on equal Froude number or centripetal acceleration. This simulation is designed to mimic full scale forces on the impeller that affect the rags coming into contact with the impeller leading edge. A square vessel was used to preserve geometric similarity with the following dimensions (Table 1). Rags were simulated using 2-4cm lengths of string which were allowed to soak in water until wetted out. The ARI2 was compared to a standard hydrofoil, 3LS39, in same vessel and experimental conditions.

Table 1. Full scale application parameters used to obtain lab scale testing parameters.

Parameter	Full Scale	Lab Scale
T - Square [m]	15.2	0.9
D/T	0.2	0.2
Z/T	0.4	0.4
Fr [-]	0.035	0.035
N [RPS]	0.33	1.36

The AR2 was observed to not collect any rags during operation however, the hydrofoil immediately collected rags as soon as the rags were suspended and the circulation loops established (Figure 8). The duration of the test for each impeller was 30 to 60 minutes. Currently, production ARI2 impellers in the field have experienced no ragging issues.

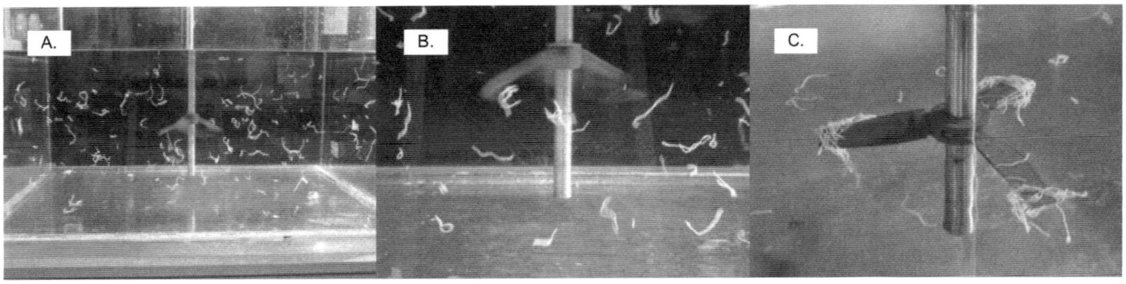

Figure 8. Ragging Test*: *(A)* ARI2 - running; *(B)* ARI2 close up; *(C)* 3LS39 (*any* typical hydrofoil) - stopped. *All tests - identical impeller speed and ragging conditions

4. CONCLUSION

The ARI2, which is used predominantly in waste water treatment mixing applications, is efficient, has the ability to resist fibrous buildup, and has been designed with cost-effective manufacturing techniques in mind. The ARI2 has comparable efficiency in terms of flow per unit cost to its sister impeller the XELSC-2200 but with a lower power number. As blending performance is dependent on unit power input, this impeller will inject less power which will increase blending time. However, this process is not sensitive to blend time and is instead sensitive to flow and tank circulation. The residence times of the tanks are large enough that blend time is not significant.

5. REFERENCES

[1] Fort I., 2011. "On Hydraulic Efficiency of Pitched Blade Impellers," Chem. Eng. R&D, vol. 89,pp. 611-615.

[2] Brown, D., 2010. "Mixer Performance Characteristics: Impeller and Process Efficiency," *Mixing XXII*, (Victoria, BC 22 June), BHR Group Limited.

(40c) Mixing Performance of the Novel Kdm Impeller in Small Laboratory-Scale Systems

Zhao Yu, Engineering & Process Sciences, The Dow Chemical Company, Midland, MI, Richard F. Cope, The Dow Chemical Co., Midland, MI and Kishore Kar, Fluid Mechanics and Mixing, The Dow Chemical Company, Midland, MI

Since small (<100 ml) lab-scale systems commonly encounter significant mixing challenges, researchers from the Engineering and Process Sciences Laboratory of The Dow Chemical Co. developed and patented a novel KDM impeller for overhead mixing within such unbaffled systems. This impeller consists of multiple twisted ribbon elements mounted on a rotating shaft. Through laboratory experiments and computational fluid dynamics (CFD) simulations, the performance of the KDM impeller was compared to that of vendor-offered impellers in three different lab-scale mixing systems. In fluid viscosities ranging from 5,000-30,000 cP, the KDM impeller produced the desired mixing in significantly less time than each commercially-offered impeller system.

Experiments and simulations characterized KDM impeller operations and yielded the optimum ratios of KDM diameter-to-vial diameter (D/T), liquid submergence-to-KDM diameter (s/D), and off-bottom clearance-to-KDM diameter (c/D). The optimal speed at which the KDM impeller rotates depends on the fluid system of interest. The design of the KDM impeller was refined for the applications described in this work by optimizing the length-to-diameter ratio (L/D) of the impeller elements, as well as the number of repeated KDM elements so as to meet the recommended values of s/D, c/D, and L/D. Other crucial features of the optimized KDM design included twisting all elements in the same direction (be it clockwise or counter-clockwise), and the 90 degree off-set angle between adjacent elements. The KDM impeller is well suited to laboratory and mini-plant scales, but its high relative mass could make it impractical at some larger scales, depending on fluids of interest, and materials of construction. Identifying the size at which the KDM becomes impractical was beyond the current effort.

(40d) The Simplest Stirred TANK: Structure and Mixing Dynamics in Eccentricaly Agitated Vessels

David Bulnes-Abundis and Mario M. Alvarez, Centro de Biotecnología-FEMSA, Tecnológico de Monterrey at Monterrey, Monterrey, Mexico

Eccentric agitation in stirred tank systems has been proposed as a cost effective way to improve mixing performance in laminar and turbulent systems. For example, at some eccentricity values, segregated regions in stirred tanks are drastically minimized, and flow separatrixes are eliminated. The flow structure, and therefore the mixing topology, of eccentrically agitated systems are far more complicated than those observed in concentrically stirred tanks. Manifolds in eccentric flows are highly asymmetrical, and dynamically more complex that concentric flows. Indeed, it is not intuitive to predict, *a priori*, the flow structure in an eccentric system. In this contribution, we use planar Laser Induced Fluorescent experiments (p-LIF) to visualize the flow structure originated by tracer injections in eccentrically agitated stirred tanks in the laminar regime ($Re \approx 40$). Eccentricity is defined in terms of the parameter E; where E=distance between the center line and the axes of rotation divided by the tank radius. Results from LIF experiments conducted at E=0.0 (concentric case), E=0.13, E=0.21, E=0.42, and E=0.50 are presented, and a mixing time is calculated based on image analysis techniques.

In addition, we use CFD to describe the short term mixing dynamics of tracers injected at different points in eccentrically agitated tanks. In the mixing literature, there is abundant discussion on the relevance of the location of injection points for the achievement of rapid dispersion. However, most of the previous work has been circumscribed to conventional stirred tank geometries, namely stirred tanks agitated concentrically, with typically used impellers. In this contribution, we extend the discussion on the relevance of a proper selection of injection points to eccentrically agitated systems. The manifold of the flow determines the fate of any tracer injected into the system. Therefore, a detailed calculation of this manifold is required to precisely diagnose the effectiveness of an injection point. Through CFD studies of the short term evolution/dispersion of tracer injections, composed of multiple particles, we compare the rate and extent of mixing at four different injection points and three eccentricities values in stirred tanks operated at $Re \approx 4000$. The rate of mixing is faster in the vicinity of E=0.42. Counter-intuitively, at E=0.50 the mixing performance is inferior than the one observed at E=0.42.

(40e) Preliminary Work On an Expert System for Mixed Reactor Design

Gary K. Patterson, Chemical and Biological Engineering Dept., Missouri University of Science and Technology, Rolla, MO

Choices of the method needed to design a reactor with mixing as an important aspect of its operation are frequently difficult to make and false starts are not uncommon. This presentation will outline the start of an effort to develop a system to aid the designer in determining which style of reactor to propose for the job and to lead that person in finding a design method which best fits the needed reactor and its characteristics. Such an approach will help the designer avoid choosing a reactor type based on a preceived bias based on previous experience and/or company history. The first aspect of the system is the determination of how much mixing rate affects the reaction and its yield of desired product. Then the needed mixing rate is determined and various choices for achieving that rate are explored via a set of approximate methods for each type of reactor deemed likely. Types of reactors included in this part of the procedure are single impeller mixed vessel, multiple impeller mixed vessel (with aspect ratio variations), impeller type and location variations, gas sparging method and location, pipe reactor with and without internal mixing (static or dynamic), stirring intensity based on solids suspension, stirring or static mixing limits based on needed gas dispersion and flooding limits. Liquid-liquid reactors (such as suspension polymerization) are not yet considered. Finally, the designer will be led to the most likely design method for the reactor type selected.

(40f) Pressure Drop Measurements and 3D CFD Modelling in the Netmix® Reactor

Carlos M. Fonte, M. Enis Leblebici, Madalena M. Dias and José Carlos B. Lopes, LSRE - Laboratory of Separation and Reaction Engineering - Faculty of Engineering - University of Porto, Porto, Portugal

The NETmix® Reactor is a new technology consisting of a network of mixing chambers interconnected by transport channels [1]. Networks are generated by the repetition of unit cells where each unit cell consists of one chamber and two inlet and two outlet channels oriented at a 45° angle from the main flow direction. Above a critical channel Reynolds number, the system evolves to a self-sustained oscillatory laminar flow regime inside the mixing chambers inducing local strong laminar mixing. This occurs due to the geometric characteristics of NETmix®. A network model was developed to describe and predict the behaviour and performance of NETmix® [2]. From the point of view of modelling, it was shown that chambers can be assumed to behave as perfectly mixing zones and the channels as plug flow perfect segregation zones. The NETmix® Reactor can therefore be understood, along the main flow direction, as a plug flow reactor with maximum mixing. The mixing degree was defined and quantified [2, 3] showing that mixing can be controlled effectively and efficiently making it particularly suited for complex and fast kinetics reactions. However, the performance of the NETmix® reactor, in terms of energy requirements, was yet to be studied.

In this work, pressure drop was measured experimentally for one NETmix® configuration with spherical chambers and two configurations with cylindrical chambers and for several channel Reynolds numbers in the laminar regime. From the experimental data, a pressure drop model was developed which takes into account the friction effects within the channels and the inertial effects due to the sudden expansions and contractions in the interconnections between chambers and channels. The model has a single adjustable parameter that is only dependent on the geometric configuration of the network. The dynamic measurement of pressure drop was also used to evaluate the mixing dynamics in the NETmix® chambers and above the critical Reynolds number, the natural oscillation frequency was quantified. The Z factor and power number were determined to compare the performance of different NETmix® configurations with other existing static mixers. Furthermore, a three-dimensional Computational Fluid Dynamic (CFD) transport model of a representative portion of the NETmix® was also developed and validated by comparing the experimental pressure drop data with the CFD simulation results.

NETmix®'s mixing efficiency was previously estimated. However from this work, its energetic performance was quantified and shown to be competitive with the compared existing static mixers. The new 3D CFD transport model allows the computation of transport properties and overcomes the need of experiments each time a new NETmix® configuration is designed. It also allows a better understanding of the flow patterns inside the unit cells which were not possible to visualize from previous two-dimensional CFD simulations.

[1] J.C.B. Lopes, P.E. Laranjeira, M.M. Dias, A.A. Martins, Network mixer and related mixing process. PCT/IB2005/000647, February 2005. European Patent EP172643 B1, October 2008.

[2] P.E. Laranjeira, A.A. Martins, J.C.B. Lopes, M.M. Dias, NETmix®, a new type of static mixer: Modeling, simulation, macromixing, and micromixing characterization, AIChE Journal, 55 (2009) 2226-2243.

[3] P.J. Gomes, C.P. Fonte, R.J. Santos, M.M. Dias, J.C.B. Lopes, Experimental and numerical characterization and quantification of mixing in a NETmix® reactor, in: ECCE 7 - 7th European Congress of Chemical Engineering & CHISA 2010- 19th internation Congress and Process Engineering, Process Engineerig Publisher, Prahe, Prague, Czech Republic, 2010.

(97a) Applied Filtered Density Function

Peyman Givi, Mechanical Engineering and Materials Science, University of Pittsburgh, Pittsburgh, PA

The optimal means of capturing the detailed, unsteady physics of turbulent reacting flows, as it is now widely believed, is via large eddy simulation (LES). The challenge in LES is accurate and consistent modeling of the subgrid scale (SGS) quantities. The filtered density function (FDF) methodology; including its mass weighted form, the filtered mass density function (FMDF), has proven particularly effective for this purpose. The FDF is essentially the probability density function (PDF) of the SGS quantities. Therefore, it provides all the statistical information pertaining to these quantities.

In its stand-alone form, the FDF must account for the joint statistics of all the relevant physical variables. The most sophisticated FDF closure available to-date is the frequency-velocity-scalar FMDF (FVS-FMDF), and a simpler version (VS-FMDF) which does not include the SGS frequency. Hydrodynamic closure in incompressible, non-reacting flows has been successfully achieved via the velocity-FDF (V-FDF) and the one which has been utilized the most only considers the scalar field (S-FDF and S-FMDF). This is the most elementary form of FDF when it was first introduced. Since then, the FDF has experienced widespread usage, and is now regarded as one of the most effective and popular means of LES worldwide. Some of the most noticeable contributions in FDF by others are in its basic implementation, fine-tuning of its sub-closures, and its validation via laboratory experiments.

In this chapter, a review is presented of modern advances in FDF modeling and simulation. These are primarily the development of new sub-closures in FDF transport, and construction of efficient solvers for this transport. With these developments, it is now possible to conduct LES of turbulent reactive flows in complex configurations. Examples are provided of recent FDF-based simulations with some speculation of the type of complex flows to be simulated within the near future.

(97b) Emerging Concepts in the Computational Modeling of Mixing and Combustion

Venkat Raman, Department of Aerospace Engineering & Engineering Mechanics, University of Texas at Austin, Austin, TX

The last decade has seen an explosive growth in computational power, which has transformed the modeling of turbulent reacting flows. With the availability of vast computational resources, model-based design of turbulent reactors is now seen as a feasible goal. In this talk, progress in the modeling complex multiscale processes will be first discussed. In particular, the evolution of large eddy simulation (LES) methodology as a predictive computational tool for reactive flows will be presented. To enable computational design, it is important to build confidence in the models. In the second part of the talk, the integration of experiments and modeling to quantify uncertainties will be discussed. An important aspect of this integration is the design of experiments that provide the most information for building model confidence. For this purpose, a new generation of computational tools based on information theory is being developed. A few of these emerging concepts including Bayesian analyses of chemical kinetics and adjoint analyses of reacting flows will be presented.

(97c) Advanced Laser Diagnostics Measurements for Reactive Mixing Flows

Robert Lucht, Mechanical Engineering, Purdue University, West Lafayette, IN

The development and application of advanced laser diagnostic techniques for reactive mixing flows are described. Techniques to be discussed include high-data-rate planar imaging techniques such as OH planar laser-induced fluorescence (PLIF). We discuss the acquisition of OH PLIF images at data rates of 5 kHz. Planar imaging techniques for measuring the extent of molecular mixing in nonreacting flows are also discussed. Coherent anti-Stokes Raman scattering measurements of reactive mixing processes are also discussed. Dual-pump CARS is used for the simultaneous measurement of temperature and species concentrations. Recent measurements of concentration and temperature fields using dual-pump H_2/N_2 CARS for a reactive jet in crossflow in a high-pressure combustion test rig are discussed. We also discuss the use of ultrafast laser systems for measurements in reactive mixing flows. We have demonstrated the use of ultrafast laser systems for 5 kHz temperature measurements in flames using chirped-probe-pulse femtosecond CARS. We also present the results of recent femtosecond CARS measurements in turbulent nonpremixed jet flames, and discuss the potential for concentration measurements using this technique.

(97d) Self-Assembling Process of Flash Nanoprecipitation in a Multi-Inlet Vortex Mixer

Ying Liu, Chemical Engineering, University of Illinois at Chicago, Chicago, IL, Robert Prud'homme, Chemical Engineering, Princeton University, Princeton, NJ and Rodney O. Fox, Chemical & Biological Engineering, Iowa State University, Ames, IA

We present an experimental and numerical study of selfassembled polymeric nanoparticles in the process of flash nanoprecipitation using a multi-inlet vortex mixer (MIVM). The four-stream MIVM allows control of both the supersaturation and the final solvent quality by varying stream velocities. The design also enables the separation of reactive components prior to mixing. It is critical for the nanoprecipitation process to have a short mixing time, so that the solvent replacement starts homogeneously in the reactor. The properties of the nanoparticles depend on the competitive kinetics of polymer aggregation and organic solute nucleation and growth. We report the existence of a threshold Reynolds number (of 1600) over which nanoparticle sizes become independent of mixing. A similar value of the threshold Reynolds number is confirmed by numerical simulation along with experimental study of competitive reactions (the socalled "Bourne reactions") in the MIVM. Flow patterns in the MIVM at various Reynolds number are microscopically visualized by mixing iron nitrate ($Fe(NO_3)_3$) and potassium thiocyanate (KSCN) to precipitate $Fe(SCN)_x^{(3-x)}$.

(97e) Clathrate Hydrate Formation in Emulsions Subjected to Shear Flow

Genti Zylyftari[1], Prasad U Karanjkar[2], Amit Ahuja[2] and **Jeffrey F Morris**[3], (1)Levich Institute, City College of New York, New York, NY, (2)Chemical Engineering, Levich Institute, City College of New York, New York, NY, (3)Levich Institute and Chemical Engineering, Levich Institute, City College of New York, New York, NY

Clathrate hydrate formation is a phase transition from fluid to solid associated with forming cages of host (water) molecules around a guest molecule, typically of hydrocarbon form, with examples including methane, propane, and cyclopentane. The last of these, cyclopentane, forms hydrates at atmospheric pressure and temperatures above the freezing point of water; these features make it rather convenient for study of the influence of the hydrate-forming reaction on the properties of an emulsion containing hydrate formers in the organic phase; we study cyclopentane hydrate formation in a water-in-oil emulsion stabilized by surfactants, focusing on sodium mooleate (an oil soluble surfactant). Our interest is in the rheology of the mixtures, and how bulk flow effects this reaction process.

Hydrate formation is relatively unusual in the reaction generating a solid phase at a fluid-fluid boundary, and this spatial localization of the reaction will be a focus in our considerations here. In particular, it wil be shown that the formation at an interface results in a pronounced influence of surface-active agents upon the reaction rate and the resulting morphology of the reactant crystals. We will describe novel hollow conical crystals and their relationship to the surfactant loading; additionally, the influence on morphology of the cystalline stucture by the surfactant will be analyzed, and correlated to observations of the evolution of rheological properties as the reaction to form hydrate progresses. An interesting feature in this system is the role of flow-induced stresses. We observe that the reaction can be shutdown by mass transfer limitations, as the hydrate shell formed at the interface isolates the water from the organic phase. As a consequence, the role of flow is not simple mixing, but also involves fracturing of shells of hydrate around water drops allowing surface wetting of existing hydrate crystals. Reaction rate jumps suddenly when this phenomenon occurs, as evidenced by the rapid release of energy associated with the exothermic hydrate formation.

(97f) Reactive Infiltration Instabilities in Fractured and Porous Rocks

Anthony J.C. Ladd, Department of Chemical Engineering, University of Florida, Gainesville, FL and Piotr Szymczak, Department of Physics, University of Warsaw, Warsaw, Poland

Reactive infiltration instabilities occur in a wide range of geophysical and geotechnical systems. The simplest such instability occurs when fluid flows between two soluble plates, which is an idealized model of fractured limestone. Even when the initial aperture is uniform at the nanoscale, an instability in the reaction front develops leading to the formation of pronounced solutional channels or "wormholes". We have suggested that this instability may explain the onset of large underground caves systems, by allowing a much deeper penetration of reactant than is possible by uniform opening of the fracture [1].

Reactive infiltration in porous media has been investigated experimentally[2], numerically[3], and theoretically [4-6]. The predictions of the linear stability analysis vary considerably depending on the underlying model assumptions. In this talk I will outline how these different theoretical approaches can be viewed as limiting cases of a more general theory. Our results span the range of geologically feasible Peclet and Damkohler numbers [7-8]. In particular we show that the convective limit [5,6] is singular, with a wavelength selection and growth rate that is very sensitive to small amounts of diffusion. We also show that the opposite limit of diffusion-dominated dissolution [4] spans a much smaller parameter range than has been supposed.

In this talk I will briefly summarize some of our most important results and discuss their relevance to geophysical systems.

[1] Szymczak, P. & Ladd, A. J. C. 2011. The initial stages of cave formation: Beyond the

one-dimensional paradigm. Earth Planet. Sci. Lett. 301, 424-432.

[2] Fredd, C. N. & Fogler, H. S. 1998. Influence of transport and reaction on wormhole formation

in porous media. AIChE J. 44, 1933-1949.

[3] Hanna, R. B. & Rajaram, H. 1998. Influence of aperture variability on dissolutional growth

of fissures in karst formations. Water Resourc. Res. 34, 2843-2853.

[4] Chadam, D., Hoff, D., Merino, E., Ortoleva, P. & Sen, A. 1986. Reactive infiltration

instabilities. J. Appl. Math. 36, 207-221.

[5] Sherwood, J. D. 1987. Stability of a plane reaction front in a porous medium. Chem. Eng. Sci.

42, 1823-1829.

[6] Hinch, E. J. & Bhatt, B. S. 1990. Stability of an acid front moving through porous rock. J.

Fluid Mech. 212, 279-288.
[7] Szymczak, P. & Ladd, A. J. C. 2012. Reactive-infiltration instabilities in rocks. Fracture dissolution. J. Fluid Mech. To be Published.

[8] Szymczak, P. & Ladd, A. J. C. 2012. Reactive-infiltration instabilities in rocks. Dissolution of a porous matrix. In Preparation.

(97g) Reducer Performance of Chemical Looping Reactors: Mixing and Reaction Effects

L. - S. Fan, Liang Zeng, Andrew Tong and Qiang Zhou, William G. Lowrie Department of Chemical and Biomolecular Engineering, The Ohio State University, Columbus, OH

With the pressing need for clean, efficient, and cost-effective energy sources, the chemical looping strategy has evolved as a promising alternative to the traditional carbonaceous fuel conversion process. The syngas chemical looping (SCL) process co-generates electricity and hydrogen with in-situ CO_2 capture from coal derived syngas through the cyclic reduction and oxidation of an iron oxide based oxygen carrier. The SCL process features countercurrent gas-solid moving-bed reactor that converts syngas to CO_2 while reducing the oxygen carriers. The countercurrent reducer in the SCL process at the Ohio State University has been operated for a combined time of more than 300 hours in a 2.5 kW_{th} bench scale moving-bed reactor and a 25 kW_{th} subpilot scale integrated system. A syngas conversion in excess of 99.5% and an oxygen carrier conversion of nearly 50% have been obtained. The thermodynamic analysis indicates that the countercurrent moving-bed reducer offers better gas and solids conversion behavior, compared to the fluidized-bed reducer. In this study, effects of the mixing and the reaction in the reducer of a chemical looping reactor system operated in the two distinct flow patterns represented by these two types of reactors, i.e., countercurrent moving-bed and the fluidized-bed, are described in terms of experimental results and reactor modeling. Further, the importance of the thermodynamic properties of the oxygen carrier in the reducer performance is highlighted in this study.

(98a) Single Pass Drop Size Distribution Produced by an in-Line Rotor-Stator Mixer

Paul E. Rueger[1], B. N. Murthy[1], Kenneth T. Kiger[2] and **Richard V. Calabrese**[1], (1)Chemical & Biomolecular Engineering, University of Maryland, College Park, MD, (2)Mechanical Engineering, University of Maryland, College Park, MD

Continuous or in-line high shear mixers are broadly employed in chemical processes to produce emulsions and liquid-liquid dispersions. Despite their widespread use, there is little fundamental basis to theoretically predict or experimentally assess their performance. As a result, process development, scale-up and operation are often by trial and error, leading to higher processing costs, start-up problems and lost time to market. In order to develop a fundamental understanding, at least on a mechanistic basis, of how drop breakup affects liquid-liquid dispersion performance, we have measured power draw and single pass drop size distributions (DSD) exiting a IKA Labor Pilot 2000/4 in-line rotor-stator mixer. This slot and tooth device can accommodate up to 3 stages of generators. We have examined medium (2 rows of teeth per stage), fine (3 rows per stage) and ultra fine (4 rows per stage) generators.

A continuous, turbulent water phase is fed to the mixer via a progressive cavity pump. At time zero, a single large oil drop is injected into the device and the resulting daughter DSD is measured at the exit using our previously reported Phase Doppler Anemometry (PDA) technique. Since PDA yields the time history of the exiting drop population, or residence time distribution (RTD), bivariate statistics of drop size and residence time can be obtained from several parent drop trials. Power draw is measured via a purpose build and fully integrated torque meter. The background load needed to overcome bearing friction and other system losses is measured using "blank" stages and subtracted from the total load to determine the Power number due to rotor rotation. Power due to flow losses is estimated from flow rate and pressure drop measurements.

Power draw has been measured for the various generators over a broad range of flow rate and rotor speed for 1, 2 1nd 3 stages. Systematic experiments have been performed with a viscous drop phase to study the effect of flow rate, rotor speed generator geometry, and number of stages on the resulting DSD and RTD. Results indicate that flow rate is not as influential as geometry and rotor speed, especially for multiple stages. Power measurements are used to aid in interpretation of the DSD data. The results will be discussed in detail, as well as implications to process performance.

The Effect of Tip Chord Angle on Hydrofoil Efficiency at Suspending Solids in a Stirred Tank

Presented at AIChE Annual Meeting 2012
Pittsburgh, Pennsylvania
Session #98 North American Mixing Forum
Mixing in Multi-Phase Systems
Submitted Paper No. 281324

Richard O. Kehn
Research & Development
SPX Flow Technology (LIGHTNIN)
Rochester, New York, USA

Joseph Kontur (presenting author)
Applications Engineering
SPX Flow Technology (LIGHTNIN)
Rochester, New York, USA

ABSTRACT

Solids suspension mixing applications exist throughout the chemical industry. The hydrofoil impeller is the impeller of choice for such applications due to its axial flow pattern and low torque requirements. Although often lumped together as one style of impeller, regardless of specifics, the exact combination of hub chord angle, blade twist and thus tip chord angle leads to different torque requirements for a given suspension application. Minimizing both power and torque leads to an optimized mixer design.

This study compares the performance of the LIGHTNIN A510 hydrofoil impeller in a solids suspension application where the tip chord angle is varied from 12 degrees to 32 degrees. Two different solid sizes were modeled, 600 and 250 micron glass beads at a solids weight percent of 2%, with the liquid being water. A 445 mm diameter tank with a dished bottom and four 38 mm wall baffles was used for the test set up. Three different sized diameter A510 impellers were tested. The just suspended speed was determined visually in each case and the data was used to compare the performance of each on a power and torque basis. The results were then scaled up using geometric similarity and a constant power per unit mass basis in order to compare relative mixer costs (both capital and operating) for a 43 m^3 tank. Although only a qualitative study due to the small sample size, this study demonstrates how small differences in hydrofoil design can lead to a more optimized mixer selection in the full scale.

A510 HYDROFOIL IMPELLER DESIGN

The hydrofoil impeller was developed in the early 1980's in order to meet the demand for a more efficient mixer. Mixer efficiency can be defined in various ways and often includes comparing variables such as impeller power, flow and torque. For solids suspension applications, the impeller must impart energy into the fluid creating a turbulent flow field near the bottom of the vessel. This action causes the solids to lift from the vessel base and thus be suspended (Handbook of Industrial Mixing p. 548). Hydrofoil impellers are typically used for such applications since they impart a highly axial flow pattern into the fluid, thus creating a high level of turbulence at the very bottom of the vessel. In addition, the shape of the blades reduces the power draw, which reduces the torque requirements for these impellers over impellers such as the pitched bladed turbine and radial flow turbine.

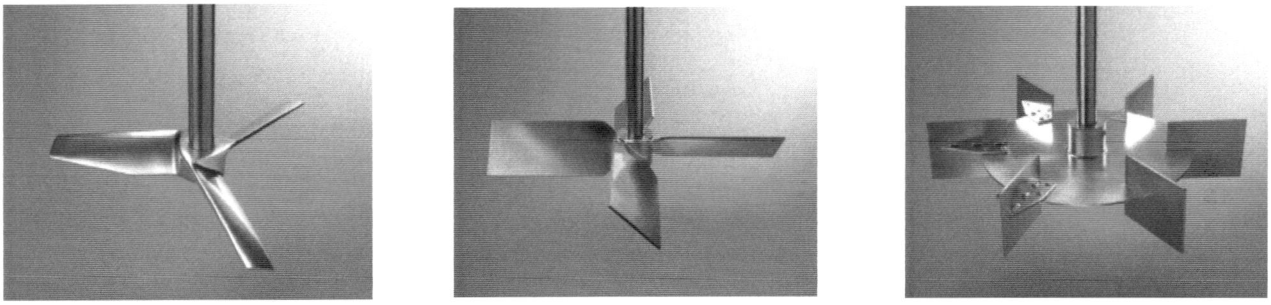

Figure 1 – A310 Hydrofoil vs. A200 Pitched Bladed Turbine vs. R100 Rushton Turbine

Since a radial flow impeller and an axial flow impeller direct flow differently in the tank, comparing both on a flow basis for a solids suspension application is not appropriate. However comparing their performance with regards to power and torque is applicable. Power and torque are calculated using the following equations:

$$Tq = \frac{P}{N}$$

$$P = \rho N_p N^3 D^5$$

The LIGHTNIN A510 is a three bladed hydrofoil impeller that can be designed over a range of tip chord angles. Its design is based on the successful A310 impeller that has served various industries for over thirty years. Both impellers are shown in Figure 2.

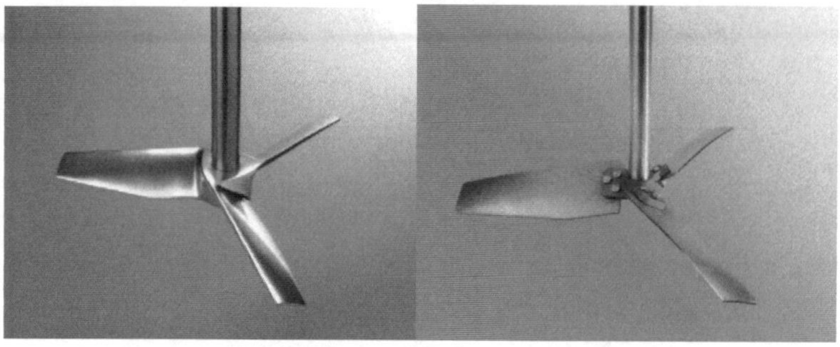

Figure 2 – A310 vs. A510

Laser Doppler Velocimetry (LDV) was used to optimize the blade shape of the A310 and A510. Both use similar blade shapes, with the angle of the blade varying from the hub to the tip. For the A510, there are three tip chord angle available, 12, 22 and 32 degrees. The flow patterns established by each design are similar to one another as shown in Figure 3 below.

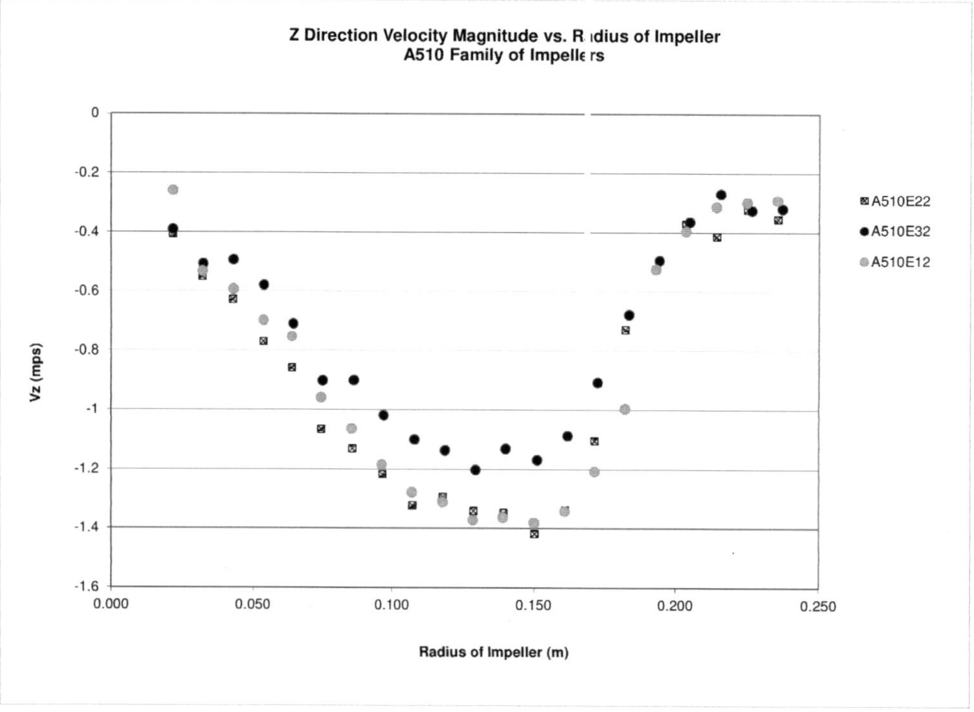

Figure 3: LDV Velocity Magnitude Comparison for A510

The relative power number comparison for each A510 impellers is shown in Table 1 below.

Table 1 – Relative Power Number Comparison

Tip Cord Angle	Relative N_p (compared to A310)
12°	0.5
22°	1.0
32°	1.5

From this table alone, the 12 degree A510 impeller would be the best choice for minimizing torque at equal power and diameter, if all three impellers suspend solids at equal power. However at equal power and diameter, the 12 degree impeller will also operate at the highest rotational speed, which could lead to mechanical design and impeller wear issues. Thus the study addresses these issues to determine which of the three impellers is the most efficient at suspending solids.

TEST SETUP AND METHODS

The testing was carried out in a 445 mm diameter tank with a dished bottom and four 38 mm wall baffles. The liquid height was maintained at 425 mm for all cases. The liquid used was water and glass beads were used to model the solids at a concentration of 2.0 %wt.

Three different diameter A510 impellers were tested at each tip chord angle for 600 micron glass beads. For the smaller glass beads (250 microns), only the smallest diameter A510 impellers were tested. In addition, a pitch bladed turbine impeller was also tested for both sized glass beads. Torque consumed by the impellers tested was directly measured via a torque cell and speed was measured using an optical tachometer. The setup is shown in Figure 4 and the test matrix is shown in Table 2.

Figure 4: Test Tank (Solids Being Suspended)

Table 2 – Test Matrix

Impeller Description	TCA	Off Bottom Clearance	D/T	600 Micron	250 Micron
150 mm A510	12	75 mm	0.340	X	X
150 mm A510	22	75 mm	0.340	X	X
150 mm A510	32	75 mm	0.340	X	X
150 mm A200	45	75 mm	0.340	X	X
173 mm A510	12	86 mm	0.390	X	Not Tested
173 mm A510	22	86 mm	0.390	X	Not Tested
173 mm A510	32	86 mm	0.390	X	Not Tested
193 mm A510	12	97 mm	0.440	X	Not Tested
193 mm A510	22	97 mm	0.440	X	Not Tested
193 mm A510	32	97 mm	0.440	X	Not Tested

For each test run, an off bottom clearance of half the impeller diameter (0.5D) was used. This meant that the off bottom clearance increased slightly as the D/T increased. The just suspended speed (N_{js}) was determined visually with the criteria being that no solids remained stagnant for more than two seconds on the tank bottom (Handbook of Industrial Mixing p. 556). Although the visual observations were qualitative, an effort was still made to maintain a level of consistency in the results by having the same two engineers determine N_{js} and by running the tests consecutively. The last point of suspension was always directly below the impeller at the bottom of the dish. The results for N_{js} also closely follow that predicted by the Zwietering correlation using the appropriate impeller constants (Handbook of Industrial Mixing p. 558).

TEST RESULTS

The results from the A510 testing using 600 micron Glass Beads is shown in Figure 5. The results clearly show that the 22 degree A510 yields the lowest power to achieve off bottom suspension over the other two tip chord angles. The overall power required to suspend the solids for all tip chord angles increases as the D/T deviates from the optimum, which is near 0.33. These results agree with previous studies (Ayranci et al 2012). It is important to note at the optimum D/T, the 22 and 32 degree results are close to one another (within 8% of one another) whereas the 12 degree requires 25% more power. The power required for suspension in the full scale will affect both initial mixer cost and annual operating costs as shown later in the scale up exercise.

When comparing the results on a torque basis, it is clear the 32 degree requires the most torque, which is not unexpected since its power number is the highest and thus will always operate at the lowest speed for a given power input. Torque can directly impact up front capital cost of an agitator. With higher torque requirements comes larger gearboxes, larger diameter shafts, thicker impeller blades all of which combine to create a more expensive agitator. When comparing the results on a tip speed basis, the 12 degree requires significantly higher tip speed over the 22 and 32 degree, which can adversely affect impeller wear and mixer mechanical design.

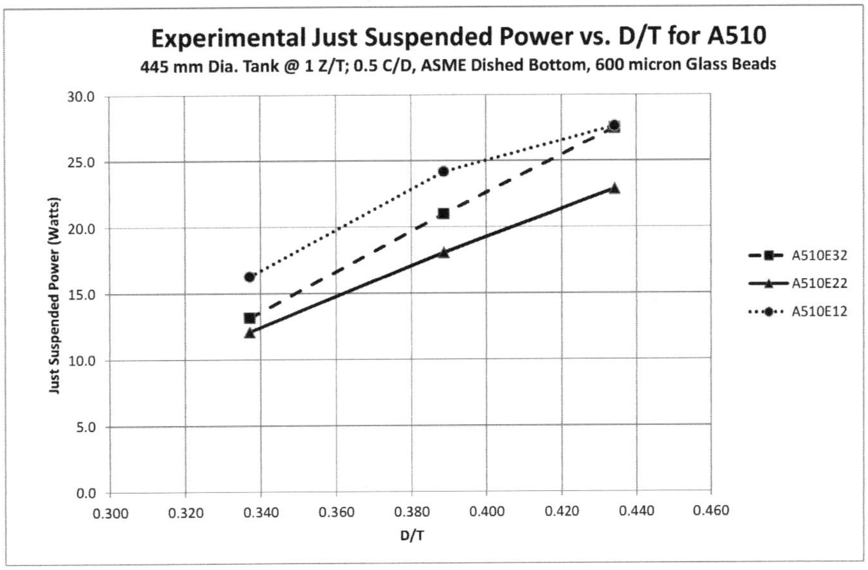

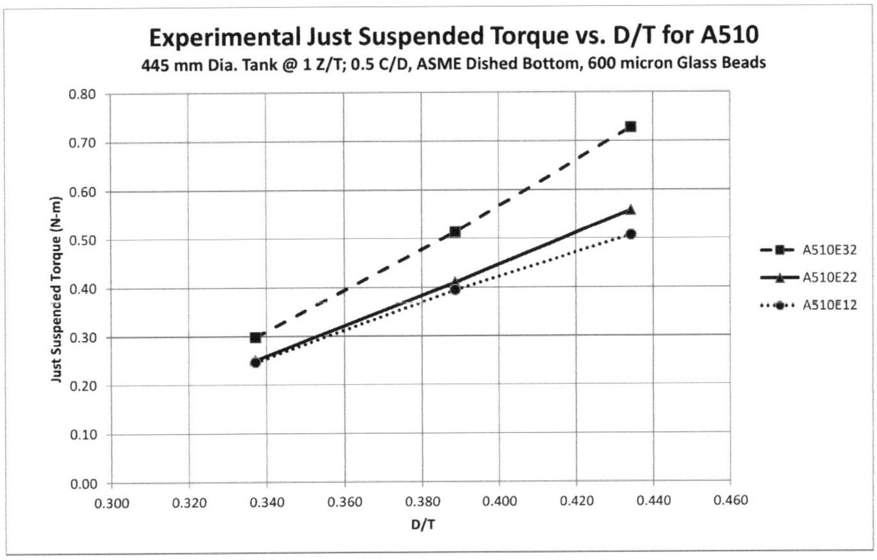

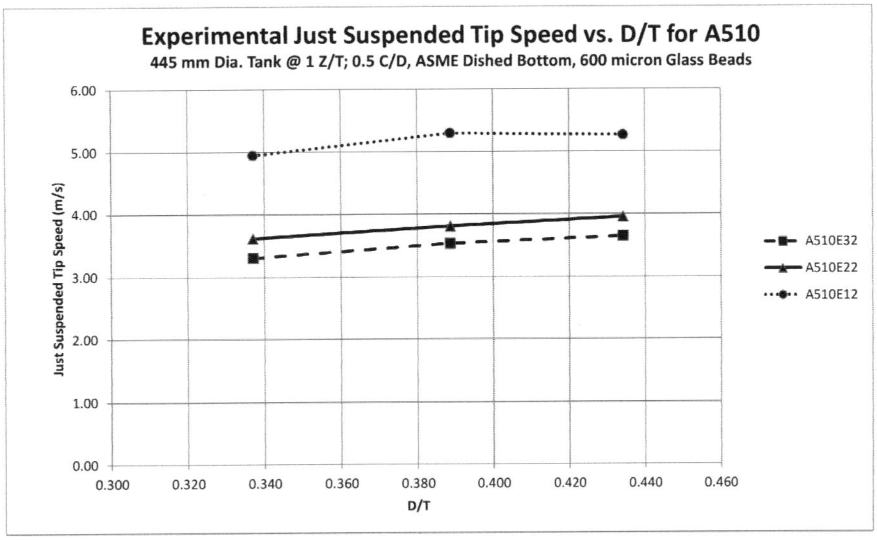

Figure 5 – A510 Experimental Results (600 Micron Glass Beads)

The test results for 250 micron glass beads are shown in Figure 6. These tests were only carried out with the near optimum D/T impellers (150 mm). The overall trends seen with 600 micron glass beads were also seen with 250 micron glass beads as well.

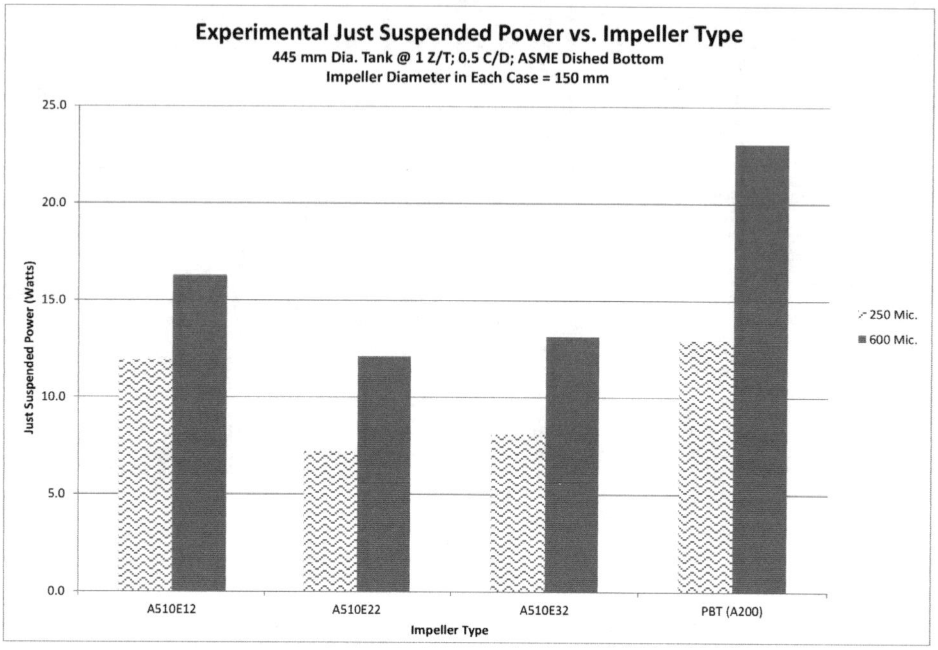

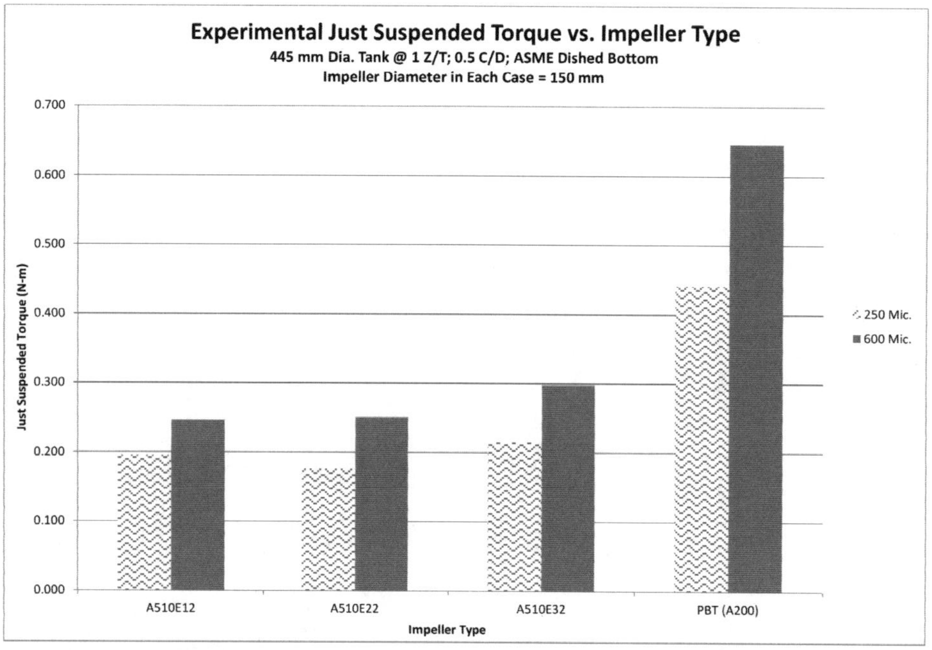

Figure 6 – A510 Experimental Results for 250 Micron Glass Beads

SCALE UP EXERCISE

Using the data for the lowest D/T (0.34) A510 impeller, the largest D/T (0.44) A510 and a Pitched Blade Turbine (PBT), those results were scaled up to a 43.5 m^3 tank. Geometric similarity was maintained between scales and the mixer selections were scaled on a 1:1 power per unit mass basis. For the gearbox sizes, one (1) is the smallest and least costly and four (4) is the largest and most costly.

Table 3 – Full Scale Cost Comparison

Impeller Type	Impeller D/T	Gearbox Size	Relative Capital Cost	Relative Operating Cost	Relative Structure Cost
A510-32	0.34	Size 1	1.00	1.08	1.18
A510-22	0.34	Size 1	1.03	1.00	1.00
A510-12	0.34	Size 2	1.38	1.37	1.40
A510-32	0.44	Size 4	1.98	2.83	2.26
A510-22	0.44	Size 3	1.64	1.97	1.67
PBT	0.34	Size 2	1.22	1.83	1.78

Based on the scale up selections and cost analysis it can be seen from the first two lines of Table 3, that both the 32 and 22 degree impeller selections result in relatively similar mixer capital costs at the optimum D/T of 0.33. When considering structural costs and operating costs, the 22 degree impeller option will yield the best overall cost for the mixer. The 12 degree selection shown in line three has a higher capital cost due to the high tip speed operation requiring a pipe shaft design to meet critical speed requirements. In addition, the 12 degree option would end up operating at an unreasonably high tip speed (10 m/s) and thus would more than likely lead to high impeller wear making it a bad option to consider for this application. The 12 degree A510 is best applied in blending applications that require low power inputs (less than 0.04 kW/m^3), which are typical in water treatment plants. Although the three impellers at a quick glance look the same, clearly the details underlying their design have an effect on the final mixer design. Therefore, careful understanding of impeller design details is required when considering one impeller over another for a given application.

The larger D/T A510 data was used to predict full scale mixer sizes and clearly shows the cost increases incurred when deviating from the optimum D/T. However, there can be additional process requirements or mechanical requirements present, which would merit the use of a larger diameter impeller (e.g. reducing tip speed to improve impeller wear, gas dispersion requirements, etc.).

Finally, the comparison with the pitched blade turbine (PBT) shows that the A510 hydrofoil is indeed a more efficient and cost effective solution for solid suspension

CONCLUSIONS

The basis of this paper was to understand the effect of tip cord angle of the A510 hydrofoil impeller on N_{js}, which then impacts, required power, torque and full scale equipment size. The findings of this study show that the 22 degree tip cord angle A510 impeller provides equivalent off bottom suspension at the lowest overall full scale equipment cost when compared to the 32 and 12 degree A510 impellers. However, the 32 degree A510 yields a similar design and thus gives the design engineer more flexibility at meeting the design criteria for a mixing application. It is important to note that although the 12 degree A510 is not suited for this application, it can yield a more economical mixer design in low power blending and light duty solids suspension applications, typically seen in water treatment. Also, the results showed that deviating to far from the optimum D/T yields a higher power and torque design, which can substantially increase the cost of the full scale equipment. Proper mixer design must include an analysis of initial cost, installation cost and operating cost in order to ensure that the lowest cost solution is arrived at. The A510 impeller system provides the design engineer with a flexible design platform for achieving these goals.

NOMENCLATURE

C/D	Impeller Off Bottom Distance to Impeller Diameter Ratio
D	Impeller Diameter
D/T	Impeller Diameter to Tank Diameter Ratio
N	Impeller Rotational Speed
N_{js}	Just Suspended Speed
N_p	Impeller Power Number
P	Impeller Power Draw
T	Tank Diameter
Tq	Impeller Torque
Z/T	Liquid Level to Tank Diameter Ratio
ρ	Liquid Density

REFERENCES

Atiemo-Obeng, V.A., Penney, W. R., Armenante, P. *Solid Liquid Mixing*, in Paul, E.L, V. Atiemo-Obeng, and S.M. Kresta, "Handbook of Industrial Mixing", Wiley, New York, 2004.

Ayranci, I., Machado, M.B., Madej, A.M., Derksen, J.J., Nobes, D.S., Kresta, S.M. *Effect of Geometry on the mechanisms for off bottom solids suspension in a stirred tank.* Chemical Engineering Science, Volume 79, 2012.

Oldshue, J., "Fluid Mixing Technology", McGraw-Hill Publications Co., New York, New York, 1983.

(98c) Measurement of Wall Jets in Stirred Tanks with Varying Baffle off-Wall Clearance and Their Relationship to Solids Suspension and Cloud Height

Jason J. Giacomelli, Research and Design, Philadelphia Mixing Solutions, Palmyra, PA and Wojciech Wyczalkowski, R&D, Philadelphia Mixing Solutions Ltd., Palmyra, PA

Two-Dimensional Particle Image Velocimetry (PIV) flow measurement technique was used to measure the wall jet produced by axial pumping impeller on baffles located in an agitated vessel. Wall Jets are the mechanism by which solids are transported from the lower region of a stirred vessel to the upper region in solids liquid mixing applications. Baffles, width T/12, were made of transparent acrylic material with the ability to be adjusted for desired off bottom and off wall clearance. The wall jet velocity profile was measured in a 0.609m (24in.) diameter vessel, with Z/T of 1, vertically along the baffle from Z/T of 0.17 to 0.8 on both the high and low pressure side. Data was used to test Bittorf & Kresta correlation (Bittorf & Kresta 2003) for baffles with off-wall and off bottom spacing. Cloud height was measured in a geometrically similar vessel using d80 of ~400μm silica sand particles with concentration varied between 10 and 40% by weight. The cloud height was correlated with tip speed and different Impeller to tank diameter (D/T = 0.3, 0.4, 0.5). Cloud height of the particle was also analyzed by comparing settling velocity of particles with jet velocity at measured cloud height.

(98d) Drawdown of Floating Solids with up-Pumping Agitators

Kevin J. Myers[1], Anand K. Pandit[1], Eric E. Janz[2] and Julian B. Fasano[3], (1)Chemical and Materials Engineering, University of Dayton, Dayton, OH, (2)Chemineer, Inc., Dayton, OH, (3)Mixer Engineering Company, Troy, OH

Solids drawdown with down-pumping agitators generally relies on the formation of a vortex that deforms the free liquid surface in the central portion of the vessel. This reliance on vortex formation requires unusual baffling to optimize performance. Common approaches include narrow baffles and partial baffles located only in the lower portion of the vessel. In both cases, reduced baffling permits tangential flow that leads to a surface vortex that pulls the solids into the liquid.

Up-pumping agitators incorporate floating solids via a combination of turbulence, bulk flow, and vortices formed at baffles near the vessel wall. Thus, standard baffling works well with this configuration. The use of standard baffling leads to a better understanding of important agitator design parameters such as power draw and mechanical forces on the impeller(s) and shaft. Additionally, the use of standard baffling for solids drawdown has the benefit of readily being applied in multipurpose vessels or vessels that have been converted from one process to another. These advantages, combined with generally lower drawdown speeds, have led to the growing popularity of solids drawdown with up-pumping impellers.

Optimal design of up-pumping agitators for floating solids incorporation requires a thorough understanding of the effect of various design parameters on the drawdown speed, torque, and power requirements. Primary design parameters include impeller type, diameter, and submergence. This work makes a thorough comparison of pitched-blade turbines and highly axial-flow hydrofoil impellers in this application. Nine impeller diameter to tank diameter ratios spanning the range of industrial application from twenty to fifty percent are considered, as well as seven submergences ranging from ten to seventy percent of the tank diameter.

Since many floating solids sink after being wetted by liquid, it is common industrially to simultaneously require drawdown of floating solids and suspension of settling solids. This leads to the use of an up-pumping impeller near the liquid surface and a down-pumping impeller near the tank base. It is uncommon to have impellers pumping in opposite directions, so little is known about the flow pattern with such an impeller configuration and there might be concern that the presence of a second impeller pumping in the opposite direction would negatively impact performance. The flow pattern generated by impellers pumping in opposite directions is addressed in this work.

Visual comparison of drawdown of floating solids with up-pumping agitators and suspension of settling solids with down-pumping agitators gives the impression that these two phenomena are very similar. To determine the extent of this similarity, comparison of the drawdown and suspension speeds is made under equivalent conditions (same impeller type and diameter, same solid, same solid mass, same density difference between solid and liquid, and impeller submergence for drawdown equal to impeller off-bottom clearance for suspension).

Limited scale-up experiments are made to determine the scale-up exponent for solids drawdown with up-pumping agitators. Knowledge of the scale-up exponent can be indicative of the mechanism of the solids drawdown with up-pumping agitators that appears to be different than the mechanism of solids drawdown with down-pumping impellers.

(98e) Modelling of Turbulence in Agitated Vessels Operating At "Just Suspension Speed"

Richard K. Grenville[1], Minye Liu[1] and Alvin W. Nienow[2], (1)DuPont Engineering Technology, Wilmington, DE, (2)Birmingham University, Birmingham, UK

Just suspension impeller speed, Njs, is an important design parameter for agitators. It is the optimum speed when suspending and dissolving particles.

Hydrofoil impellers achieve Njs for half of the power input of a pitched blade turbine of the same diameter (Grenville et al. 2010) but the maximum Turbulent Kinetic Energy Dissipation Rate (TKEDR) generated by a hydrofoil is double that generated by a pitched blade turbine at the same vessel averaged power input per mass (Grenville and Brown 2012). Operating at Njs, the maximum TKEDR generated by the two impellers will be approximately equal.

In this paper we have calculated Njs for three impellers; a hydrofoil and pitched blade and Rushton turbines. CFD simulations of the single-phase flow and turbulence quantities have been calculated, at the blades and at the vessel base, in order to explain the difference in the impellers' power requirements when operating at Njs.

Recommendations regarding impeller selection for suspension and mass transfer will be discussed.

Effect of Particle Size and Shape On Just Suspended Impeller Speed

Several popular correlations for the minimum suspension speed exist for round or near round particles. In this work we looked at other fabricated particle shapes: rods and plates. The particles were larger than typical and this led us to reevaluate the effect of particle size in the 1000 micron/ 1 mm range. A number of interesting and surprising effects were found.

(226a) Vortex Generation with Submerged Baffles

Jason J. Giacomelli, Research and Design, Philadelphia Mixing Solutions, Palmyra, PA and Wojciech Wyczalkowski, R&D, Philadelphia Mixing Solutions Ltd., Palmyra, PA

It is often desired in solid-liquid or gas-liquid handling applications utilizing a stirred tank to add a secondary impeller located in the upper region of the vessel and modified baffles to generate a stable vortex at the air-liquid interface to drawdown, wet, and incorporate solids or gas. The vortex for such applications must be stable and of the minimum depth of the upper impeller submergence. The stable generation of a centrally located vortex that reaches impeller about a concentrically mounted shaft in a 0.609 m diameter cylindrical vessel with submerged baffles was investigated using a series of six different types of impellers with varying D/T = 0.3, 0.4, 0.5. The geometric factors of interest were the impeller types, D/T and *baffle/impeller-liquid coverage (IBC).* The ratio of IBC to impeller diameter (IBC/D) was varied from 1 to a level where a vortex could not be identified. The impeller horizontal centerline was fixed at the top of the submerged baffle. The minimum vortex depth was correlated with Froude Number, torque/volume, power/volume and tip speed. The aforementioned conditions were observed and classified visually.

Interaction Between Fluid Mixing and Chemical Reaction -Effect of Non-Uniform Chaotic Mixing On Concentration Distribution for Periodical Reaction System-

Shunsuke Hashimoto, Yusuke Chikamochi, Akitoshi Nishimura

Division of Chemical Engineering, Graduate School of Engineering Science, Osaka University

Introduction

Mixing in stirred tank reactors in a wide variety of tank sizes and impeller shapes has been often utilized to homogenize different substances and to conduct chemical reactions in industrial chemical processes. Recently in various industrial processes, a wide range of operation for stirred tank is required depending on purposes and conditions. "Mixing time" is essential as one of index to evaluate the mixing process in stirred tank experimentally. Various techniques for measuring mixing time have been used such as coloration, decolorization, measurements of electric conductivity and temperature, and so on. Coloration is most simple technique to observe the mixing state in a whole tank, while it has a few demerits: it is difficult to detect the mixing state in a central part of stirred tank and the required time for complete mixing. Measurement of electric conductivity (1) enables us to detect delicate time variation of conductivity, but only local information for mixing near the sensor of conductivity can be obtained. Measurement of temperature has also demerits that only local information for mixing near thermocouples can be obtained, which is similar to measurement of electric conductivity. There are additional problems such as the variation of fluid property due to temperature change and thermal insulation for the measurement of temperature. It is expected that tomography (2) can compensate these demerits for other techniques efficiently in future, while its low resolution and high cost remain the key issue for the evaluation of mixing state. Decolorization enables us to observe easily the difference of mixing state at every position in stirred tank. In the case of decolorization, it is easy to determine the completion time of mixing because the termination of decolorization is directly equivalent to the completion time of mixing. Consequently, decolorization is generally used as a simple method for the measurement of mixing time. In the case of the visualization using chemical reactions such as decolorization, however, the relation between the rate of chemical reaction and that of convective mixing is quite important.

Based on general mixing theory, the completion time for mixing should be defined as the time that is required to homogenize the concentration of each fluid component all over the place in the system. In above general techniques for the measurement of mixing time, fluid is judged to mixed completely at the time that its concentration, conductivity, and temperature are homogenized in stirred tank. The properties such as concentration, conductivity, and temperature are considered to have one-to-one relationship with substance quantity of fluid component (that is, concentration). Hence, it is reasonable to use these properties as an index to evaluate mixing state. On the other hand, there are unique cases containing fluid properties that are not one-to-one relationship with concentration. For example, in the periodical vibrational system by Beloousov-Zhabotinskii (BZ) reaction, fluid has not only the property of concentration but also the properties of period and/or phase of concentration oscillations, which depend on the dynamical mode of reaction and do not have simple relationships with the concentration. In this case, it is inadequate to judge the mixing completion based on only the homogeneity of concentration of each component in stirred tank. In addition, final pattern by convective mixing would potentially depend on only the structure of flow field. For example, under low blending Reynolds number (*Re*) conditions, the presence of segregated mixing regions from the chaotic mixing regions (CMR) in the form of toroidal vortices above and below an impeller, which is called as "isolated mixing region (IMR)" (3, 4), is well known. Material exchange by diffusion is dominant at the interface between CMR and IMR. In this case, even if the decolorization (homogenization of concentration) in the whole region of stirred tank, IMR must remain there essentially. Hence, it is necessary to evaluate mixing degree based on not only the homogeneity of concentration but also various index.

The present study focused attention on reconsidering the traditional concept of mixing that was based on the homogeneity of concentration. In the present study, required time for complete mixing and mixing patterns containing "partially mixed regions" such as IMR in CMR were investigated by use of two chemical reactions: conventional decolorizing reaction (with iodine and sodium thiosulfate) and periodical oscillating reaction (BZ reaction). Based on experimental results, the body of matter that was mixed by impeller agitation and the correspondence between chemical reaction and convection were discussed with the view of "fluid-informatic" engineering. In addition, the availability of periodical reaction for the visualization of partially mixed regions was briefly discussed.

Experimental

Materials

In the present study, we used several chemical materials (research grade) as follows: research grade sodium thiosulfate ($Na_2S_2O_3$, anhydrous, mole fraction purity 0.950), potassium iodine (KI, mole fraction purity 0.995), iodine (I_2, mole fraction purity 0.998), starch (hydrosoluble), saccharose (hydrosoluble), sulfuric acid (H_2SO_4, 0.001 mol/m^3), malonic acid (CH_2 $(COOH)_2$, mole fraction purity of 0.98), sodium bromate ($NaBrO_3$), and ferroin ($[Fe(phen)_3]SO_4$, 25 mol/m^3). All of them except for $NaBrO_3$ (Sigma-Aldrich) and $[Fe(phen)_3]SO_4$ (MERCK) were obtained from the Wako Pure Chemical Industries, Ltd. and were used without further purifications. In addition, deionized water was produced using water-manufacturing equipment made by Nihon Millipore K. K.

Apparatus

The experimental apparatus in the present study consisted of the acrylic cylindrical vessel (handcrafted), two-paddled agitating impellers (handcrafted), agitating motors (HEIDON, BL-600), syringe pump (HARVARD, PHD-2000), digital video camera (NAC., MLX-ST-706), and temperature control system (AS ONE, LTB-400). The diameter of vessel (D) was 44 mm. To reduce photographic distortion and control the temperature inside vessel, the vessel was immersed into a square vessel of acrylic resin where the thermostated water was circulated. In the present study, two-bladed paddle impeller was adopted, which was the simplest among various impellers. The diameter (d), length (l), and thickness of impeller were 22 (d/D = 0.5), 4.5 (l/D = *ca.* 0.1), and 1.5 mm, respectively. One of impeller blades was fabricated so that the tracer fluid could be injected into the vessel from the tip of the impeller blade. The diameter of impeller shaft was 4 mm. The impeller was installed on the centerline at a distance (h) of 22 mm from the bottom of the vessel, which was half the height of impeller (h/D = 0.5). The working fluid was poured to the height of 44 mm in advance, which was equivalent to the diameter of vessel (H/D = 1.0). The viscosity (μ) and density (ρ) of fluid used in the present study were measured by cone-plate type viscometer (Tokyo Keiki Co., Ltd., VISCONIC ELD) and pycnometer, respectively. Based on these data, in the present study, the blending Reynolds number defined as $Re = \rho n_r d^2/\mu$ was controlled under condition of $50 < Re < 2000$. Experimental temperature was measured by a Pt resistance thermometer (Thermoprobe Inc., TL-1A). The uncertainty of temperature and viscosity were 0.06 K and 1 %, respectively.

Procedures

Decolorization

In this method, a redox reaction of I_2 with $Na_2S_2O_3$ was utilized to study a mixing state and pattern. First of all, water solution of saccharose was prepared at the concentration of 13 mass% as a working fluid. The viscosity and density of 13 mass% saccharose solution was *ca.* 2.3 mPa•s and *ca.* 1054 kg/m^3 at 278.15 K, respectively. Then, appropriate quantities of I_2 were dissolved in saccharose solution with the help of KI and this solution was black-colored by dissolution of starch (iodo-starch reaction). The necessary quantity of this solution was set in the vessel. After that, an agitating motor was operated at desired rotating speed (rpm) that was suitable for the desired *Re*. After fluid field reached a cyclostationary state in the agitated vessel, the $Na_2S_2O_3$-saccharose solution prepared at a desired concentration was injected to the vessel from the position of $r = 22$ mm and $z = 33$ mm. The injecting quantity and rate of $Na_2S_2O_3$-saccharose solution were about 1.2 times as large as the equivalent amount of I_2 and 5.0 cm^3/min., respectively, which was based on the previous report (5). The convective-mixing region was decolorized by redox reaction, while immiscible region was stood out in bold relief as poor mixing regions. Consequently, mixing patterns were obtained in the agitated vessel. The experimental images were recorded by use of digital video camera. In addition, the time needed for complete decolorization (generally regarded as the time for complete mixing) was determined.

BZ periodical reaction

BZ reaction is well known as a periodical reaction that metal-ion oxidation and reduction alternately arise in the system and then its concentration changes autonomously and periodically, which is summarized by the following chemical reaction formula:

$$5HOOCCH_2COOH + 3BrO_3^- + 3H^+ \rightarrow$$
$$3HOOCCHBrCOOH + 2HCOOH + 4CO_2 + 5H_2O \qquad (1)$$

Figure 1(a) shows the limit-cyclic closed orbit of BZ reaction where *x*- and *y*-axes stand for the concentration of Br^- and ferriin ($[Fe(phen)_3]_3^+$), respectively (6). Wherever the initial state locates in this concentration-phase plane, the trajectory of concentration always come close to the limit cycle after the transitional state. In Figure 1(a), the arrows stand for the direction of migration of concentration phase points. Figure 1(b) shows time variation of concentration of Br^-, $[Fe(phen)_3]_3^+$, and $HBrO_2$ on the limit cycle. The color of solution is dark red at initial state, while it becomes blue under the conditions of high concentration of $[Fe(phen)_3]_3^+$. Therefore, the color of solution changes from dark red to blue for a short time. In general, the periodical color variation keeps up for several hours. As shown in Table 1, the solution A and B were prepared at desired concentrations and equal volume (33 cm^3). At first, the solution B was injected into the vessel and the system was kept static. Then, the solution A was put in the vessel rapidly. Just after that, the contents were preliminarily agitated at the rotation speed of 250 rpm for fifteen minutes (pre-mixing process, $Re = 883$). Then, the agitation speed was set at a given constant value and the mixing pattern was visualized. For the first several minutes after the pre-mixing process, the transient state occurs, where anomalous spatiotemporal pattern of concentration oscillation is observed. After that, the periodical steady color variation is repeated for a few hours, while the configuration and period of stationary color variation depend on *Re* conditions. In BZ reaction experiments, *t* is defined as the time measured after the pre-mixing process.The system temperature in the vessel was kept constant at 278.15 K. The viscosity and density of agitating fluid obtained by mixing solution A and B was *ca.* 2.4 mPa•s and *ca.* 1051 kg/m^3 at 278.15 K, respectively. In this case, the color of solution periodically and alternately changed from dark red (derived from $[Fe(phen)_3]_2^+$) to blue (derived from $[Fe(phen)_3]_3^+$). The sequential digital images of color change pattern were taken with a digital video camera.

Incidentally, it has been already confirmed that with or without pre-mixing process and initial setup of agitating fluids (for example, the injecting amount and/or position of decolorizing fluid, the setting turn of fluids (for detail, please see fluids A and B in Table 1) for the BZ reaction, and so on) would not affect the experimental results obtained in the present study. That is, the mixing time and color-variation patterns have intrinsic significance in the system.

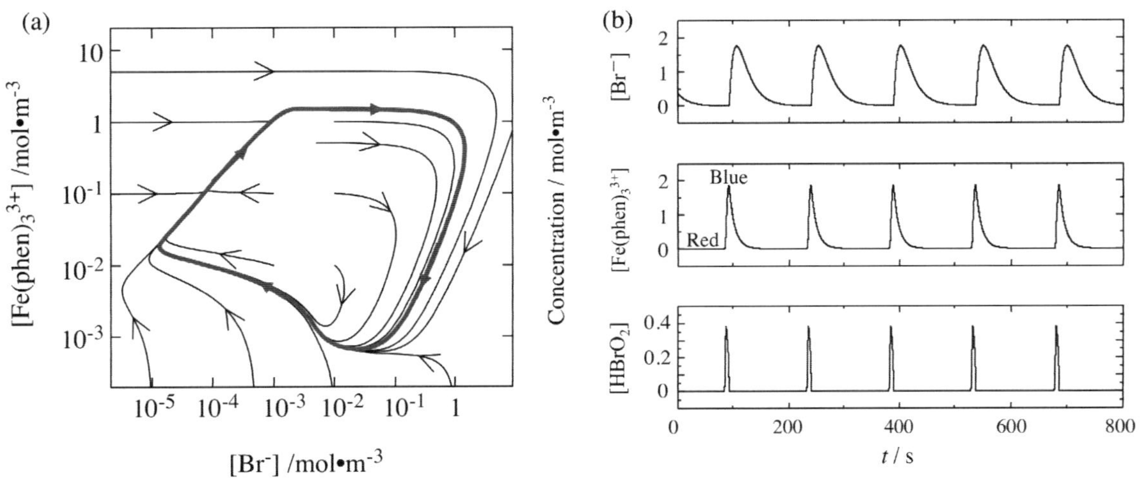

Figure 1 Limit-cycle trajectory in Oregonator model (a) (the arrows represent the direction of motion of concentration phase points.); Temporal change of concentration of three ionic species in BZ reaction: (a)Br^-, (b)$[Fe(phen)_3]_3^+$, and (c) $HBrO_2$ (b).

Table 1 Details on the concentration of aqueous solutions used in BZ reaction.

Solution A		Solution B	
$NaBrO_3$	6.00×10^2 mol/m^3	$CH_2(COOH)_2$	2.00×10^2 mol/m^3
H_2SO_4	2.00×10^2 mol/m^3	$[Fe(II)(phen)_3]^{2+}$	1.92 mol/m^3
		H_2SO_4	2.00×10^2 mol/m^3

Results and Discussion

Figure 2 shows the typical images of time variation on mixing patterns in the agitated vessels under various *Re* conditions by use of decolorizing and BZ periodical reaction. At *Re* = 117 (Figure 2(a)), as you can see in the top region of vessel, decolorizing proceeds from near the vessel wall (*t* = 22 s), and then decolorizing gradually expands near the shaft. In this case, decolorizing proceeds in the region except for the center of vortices of secondary discharged-flow (above and below an impeller), which is called as IMR (3, 4). Long after that (*t* = 459 s), decolorizing also proceeds in the bottom region of vessel. Fluid exchange between top and bottom regions is performed at only the limited region that locates near the vessel wall. That is, fluid particles are difficult to get across both top and bottom regions and consequently mixing action through both regions remains weak despite relatively high *Re*. Finally, at *t* = 1032 s, IMR region is completely decolorized, which is generally defined as the required time for complete homogeneous mixing. In the BZ reaction system (*t* = 1513 – 2247 s)), on the other hand, the following periodical color variation is repeated in the stationary state as follows: the whole system is dark red → the bottom region turns blue and dark red toroidal IMR region appears inside that → the bottom region turns dark red, while the top region turns blue and dark red toroidal IMR region appears inside that → the whole system returns dark red → ···. On the other hand, as aforementioned, the required time for complete decolorization is at most t = 1032 s at the same *Re*. That is, interestingly, the periodical steady color-variation in BZ reaction system remains even if the time required for the complete decolorization (that is, ordinary mixing time) is passed over. Under this condition (*Re* = 117), the period of color variation is *T* = 167 s. The sequential spatial color-patterns obtained in periodical variation process are similar to the transitional color patterns observed in the decolorization process until the complete decolorization.

At *Re* = 780 (Figure 2(b)), unlike the low *Re* case, decolorizing proceeds rapidly and uniformly (*t* = 10 s) in the whole system except for the small region around impeller shaft, which is generally called as "forced vortex zone". This indicates that the mixing between forced vortex zone and ambient fluid is relatively slow. At *t* = 41 s, the whole region containing forced vortex zone is completely decolorized. In BZ reaction system (t = 660 – 2660 s), the periodical steady color variation is repeated for a few hours after the transient state and the period of color variation is T = 468 s. The time for the completion of transient state (t = 41 s) is almost close to that of complete decolorization at the same Re. As you can see, the following periodical color variation is repeated in the stationary state as follows: the whole system is dark red → the system except for the small region around impeller shaft (forced vortex zone) becomes blue → the whole system becomes blue → the whole system returns dark red → ···. Hence, as in the cases with the other Re, the sequential spatial color-patterns obtained in periodical variation process are similar to the transitional color patterns observed in the decolorization process.

Figure 2(c) shows the color-pattern diagrams corresponding to these experimental images at various Re (= 117 and 780). In Figure 2(c), the region of A and A' stands for the well (chaotic) mixing region, while the region of B and B' does the poor mixing region (partially mixed region) such as toroidal IMR and forced vortex zone. Hence, the mixing rate of substances is fast in the region of A and A', while that is relatively slow in the region of B and B'. The exchange rate of substances between the well and bad mixing regions (that is, between A and B or A' and B') is relatively slow compared with the mixing rate of each one region. As a result, the partially mixed region inevitably appears and remains in the vessel. In addition, although both regions of A and A' are well mixing region and consequently the exchange of substances in each region is efficient, the exchange rate of substances between the region of A and A' is relatively slow. This is because the fluid exchange between top and bottom region in the impeller-agitated vessel is performed not near the impeller blade but through only narrow route around the stagnation point near the vessel wall, where the discharge flow from the impeller impacts (7). In the decolorizing experiment, the spatial color-patterns shown in Figure 2(c) appear at the transient state until the complete decolorization. On the other hand, in the BZ reaction experiment, these spatial color-patterns are observed steadily and periodically after the mixing of substances are completed. This may be derived from the existence of some partially mixed regions in the vessel, where the exchange rate of substances is slow. In each partially mixed region, the period and phase of color variation are synchronized and uniformalized, while they cannot be synchronized completely between other partially mixed regions because of slow exchange rate of substances. In the present study, these result in the phase gap of periodical color variation and its retention.

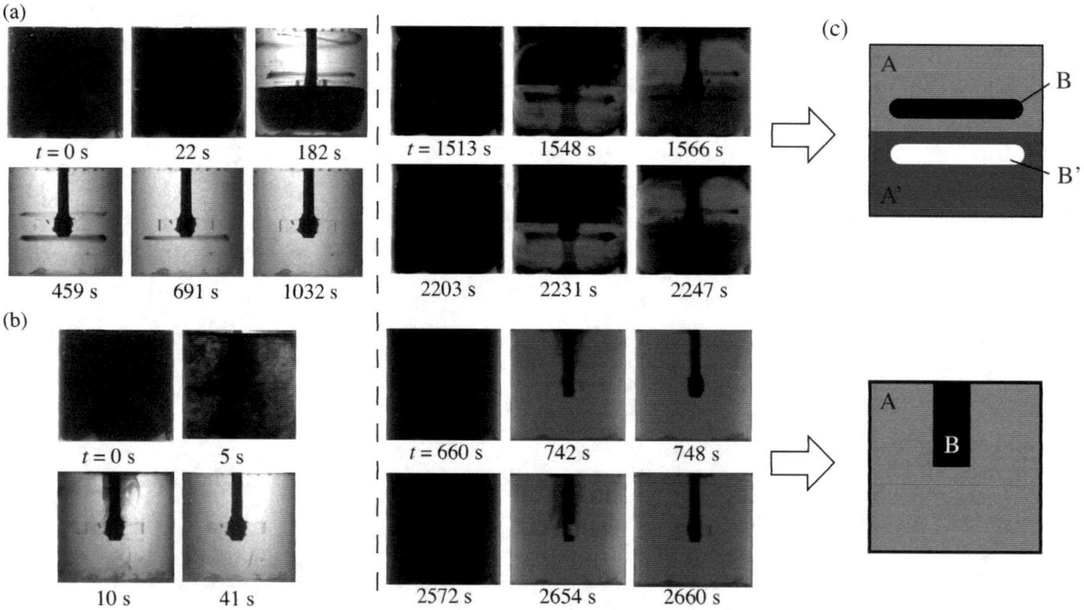

Figure 2 The experimental images obtained from the decolorizing experiment and the BZ reaction experiment; (a)Re = 117 and (b)Re = 780. Panel (c) shows the color-pattern diagrams corresponding to these experimental images.

As aforementioned, the relative speed between the exchange of substance and the synchronization of periodical oscillation of concentration dominates the probability of phase synchronization of partially mixed regions that have oscillation patterns containing phase gap. In fact, as shown in Figure 2, the area of well mixing region where the color changes in the same phase periodically becomes large as *Re* increases, which means the enhancement of convective flow. At another larger *Re* where the exchange rate of substance is extremely fast, the mutual transfer of fluid substances can occur rapidly compared with the time required for the synchronization of phase of all oscillators in the vessel. As a result, the peculiar spatiotemporal pattern of concentration oscillation can appear, where each oscillator locates independently and randomly without the synchronization of phase. In this case, the whole system looks as if the concentration did not change, in spite of the concentration change of each oscillator. Figure 3 shows the dependence of *Re* on oscillation period of BZ reaction. In fact, the concentration oscillation in the whole vessel apparently comes to a stand at $Re > 1500$. Incidentally, the experimental results using relatively high viscous fluid (*ca.* 20 mPa•s) show that the oscillation period of concentration in BZ reaction becomes 1.2 times as long as that of low viscous fluid (3 mPa•s). In addition, the phase gap between partially mixed regions at $Re = 195$ using high viscous fluid is somewhat large (*ca.* 7 s) compared with the case of low viscous fluid (4.4 s). The effect of molecular diffusion is generally potent in low viscous fluid compared with that of high viscous one. Thus, the exchange rate of substance between each adjacent concentration oscillators is relatively small because of the lack of molecular diffusion effects, and consequently the interaction among oscillators is weak. This may result in the long period of concentration oscillation. These results support that the exchange rate of substances between each adjacent concentration oscillators is one of dominant factors for the synchronization phenomena of spatial concentration oscillation in mixing system with BZ reaction.

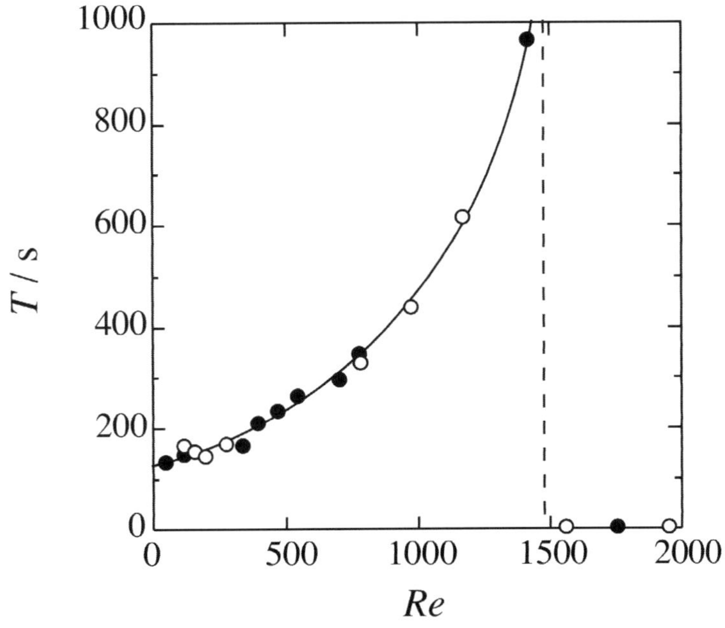

Figure 3 Relationship between *Re* and oscillation period of BZ reaction, *T*. Closed and open keys show the experimental results with and without pre-mixing, respectively.

Conclusion

In the present study, mixing time and pattern in the impeller-agitated vessel were evaluated by means of two chemical reactions: ordinary decolorizing experiment (with iodine and sodium thiosulfate) and BZ periodical reaction experiment at various *Re*. The periodical steady color-variation in BZ reaction system remains after the time required for the complete decolorization (that is, ordinary mixing time) is passed over. The sequential spatial color-patterns obtained in steady periodical variation process are similar to the transitional color patterns observed in the decolorization process until the complete decolorization. The color-patterns obtained in BZ periodical reaction are consistent with the outline of partially mixed regions where the exchange of substance is relatively slow in the vessel and they depend on *Re*. The phase of periodical concentration oscillation in each partially mixed region is shifted with one another in spite of the same period of oscillation. The relative speed between the exchange of substance and the synchronization of periodical oscillation of concentration dominates the probability of phase synchronization between partially mixed regions that have oscillation patterns containing phase gap.

References

1. Kramers, H., Baars, G. M. and W. H. Knoll, (1935). "A Comparative Study on the Rate of Mixing in Stirred Tanks," Chemical Engineering Science, 2, pp. 35-42.
2. Kaminoyama, M., S. Taguchi, R. Misumi, and K. Nishi, (2005). "Monitoring Stability of Reaction and Dispersion States in a Suspension Polymerization Reactor Using Electrical Resistance Tomography Measurements," Chemical Engineering Science, 60, pp. 5513–5518.
3. Bresler, L., T. Shinbrot, G. Metcalfe, and M. J. Ottino, (1997). "Isolated Mixing Regions: Origin, Robustness and Control," Chemical Engineering Science, 52, pp. 1623-1636.
4. Hashimoto, S., H. Ito, and Y. Inoue, (2009). "Experimental Study on Geometric Structure of Isolated Mixing Region in Impeller Agitated Vessel," Chemical Engineering Science, 64, pp. 5173-5181.
5. Nishi K., (2008). *Mixing Characteristics, The Society of Chemical Engineers, Japan Ed.*, Sankeisha, Aichi, Japan.
6. Field, R. and H. -D. Fösterling, (1986). "On the Oxybromine Chemistry Rate Constants with Cerium Ions in the Field-Körös-Noyes Mechanism of the Belousov-Zhabotinskii Reaction: The Equilibrium $HBrO_2 + BrO_3^- + H^+ \leftrightharpoons 2BrO_2^{\cdot} + H_2O$," Journal of Physical Chemistry, 90, pp. 5400-5407.
7. Inoue, Y. and S. Hashimoto, (2010). "Analysis of Mechanism of Laminar Fluid Mixing in 3-D Mixing Tank Based on Streakine Lobes," Kagaku Kogaku Ronbunshu, 36, pp. 355-365.

(226c) Mixing Time in Partially Filled Stirred Vessels

Shilan Motamedvaziri and **Piero Armenante**, Otto H. York Department of Chemical, Biological and Pharmaceutical Engineering, New Jersey Institute of Technology, Newark, NJ

Mixing time, the time required to achieve a predefined level of homogeneity of a tracer in a mechanically stirred vessel, is an important parameter to evaluate the mixing efficiency of the system. In this work, mixing time was obtained for the case in which the liquid height-to-tank diameter ratio, H/T, is less than 1, i.e., for partially filled vessels. The time, t_{95}, required to achieve a 95% level of tracer homogeneity a stirred vessel with a disk turbine with D/T=0.31, C_b/T=0.30 and for different H/T values (corresponding to different values of the impeller submergence ratio, S_b/D) was measured experimentally using a colorimetric technique coupled with image processing and predicted computationally using CFD simulations. In addition, a correlation between mixing time and t_{BR}, defined as the ratio between the liquid volume in the tank and the flow rate generated by the impeller, was developed. The experimental results show that there is a critical impeller submergence ratio S_{bc}/D below which mixing time increases suddenly and significantly (implying that the mixing efficiency drops significantly) compared the mixing time at higher submergence ratios. Above the critical impeller submergence ratio t_{95} decreases with decreasing impeller submergence ratio although the non-dimensional mixing time, $t_{95}N$, remains constant irrespective of the agitation speeds for each impeller submergence ratio. Lower values for $t_{95}N$ can be observed for lower submergence ratios. In general, good agreement between the experimental data at different S_b/D ratios, the predicted results from CFD simulations, and the proposed correlation for mixing time was obtained.

(226d) Agitator Power Draw in Continuous Operation

Mark A. Heimovitz, Ashta Chemicals, Ashtabula, OH, Kevin J. Myers, Chemical and Materials Engineering, University of Dayton, Dayton, OH, Eric E. Janz, Chemineer, Inc., Dayton, OH and Julian B. Fasano, Mixer Engineering Company, Troy, OH

Chemical and related industries often equip processes with mechanical agitators in continuous operation. Continuous systems with high incoming flow rates, such as those in water treatment applications, have experienced deviations between actual agitator power draw and that predicted with experimental data from batch studies. Occasionally, underestimating power draw can have costly consequences, such as shutting down a process for repairs. This work investigates the effects of flow on agitator power draw due to variables including impeller type and diameter, inlet diameter, and imposed flow rate for an agitated vessel in continuous operation with a bottom-centered inlet.

It has been observed that some continuous systems experience significant differences in power draw from batch systems. Three radial-flow disc impellers were tested and each experienced nearly linear increases in power draw with increasing imposed flow. The S-4 impeller, a radial-flow impeller with no disc, experienced an increase in power draw with increasing imposed flow until reaching a maximum, after which the power draw rapidly decreased to a minimum, and then gradually began to rise again. The P-4 and HE-3 impellers in the down-pumping orientation experienced increases in power draw, which were steeper for the HE-3. The P-4 in the up-pumping orientation experienced a slight increase in power draw with increasing imposed flow rate, followed by a slight decrease, whereas the up-pumping HE-3 experienced a significant power draw decrease.

General graphical correlations have been developed as the ratio of power number with flow to the power number without flow as a function of the imposed flow velocity to impeller tip speed ratio, in disagreement with literature models that predict that the power number ratio is determined by the ratio of imposed volumetric flow rate to impeller pumping rate. Only the S-4 data was correlated by the flow ratio. Some systems diverge from the correlations due to less imposed flow contacting the impeller discharge by mechanisms of bypassing and flooding. Also, some deviations may be due to data uncertainty and system temporal variance, which are difficult to quantify and should be considered in future studies.

(226e) Predicting Blend Time and Mixing Efficiency of Different Types of Impellers with Mean Age Method

Minye Liu, DuPont Engineering Technology, Wilmington, DE

The mean age method is assessed to characterize impeller mixing performance of stirred tank reactors with three types of commonly used impellers in process industries: PBT, A310, and Rushton. The spatial distribution of mean age reveals the non-uniform distribution of local blend time. The mean age frequency function is found to scale with impeller speed. The function can also be scaled with the ratio of the impeller diameter to the tank diameter but the exponent of the ratio is slightly different for each type of impellers. A blend time can be defined from the mean age frequency function and this blend time is found to agree very well with the blend time correlations in literature. This method can be used to predict blend time in a stirred tank reactor using CFD with only a small fraction of the CPU time required by the current method of solving the time dependent tracer scalar equation. With this method, CFD can be efficiently used for practical reactor analysis, design, and scale-up.

(226f) Experimental Modelling of the Free Surface Shape in an Unbaffled Tank with a Newtonian Fluid

Paul A. Mosser, **Mourad Heniche** and Louis Fradette, Dept. of Chemical Engineering, URPEI, École Polytechnique de Montréal, Montréal, QC, Canada

One of the most significant work on the prediction of the free surface shape in agitated tanks dates back to Nagata's early contribution, over 40 years ago. Despite the usefulness of Nagata's findings, more work is required in order to predict the shape of free surfaces accurately, especially in industrial conditions. The aim of this work was to evaluate the applicability of two experimental methods for determining the free surface shape from the top of a tank. A Newtonian fluid of low viscosity was used to work in the turbulent flow regime. The first method relies on image processing and grey scales to determine both the shape and the depth of the vortex. The second one makes use of a laser beam to determine analogous information. The third method, used as a reference, is based on lateral observations of a transparent tank and on video recordings. To assess the performance of each method, we compare their accuracy and repeatability. Our results indicate that both methods (grey shades and laser beam) can provide reliable results for obtaining the shape and depth of the vortex. However, more variability was observed with the grey shades method. Additionally, the methods tested can provide quantitative measurements leading to the modelling of the shape of the vortex. The investigation methods presented herein will be helpful to determine vortex shapes and depths in industrial mixing conditions where transparent tanks and fluids are not used.

Thirty Four Years of Teaching Process Mixing to Undergraduates. with Some Learnings

Fluid Mixing is rarely taught to undergraduates even as part of a fluids course. Most graduates acknowledge once in industry the topic of mixing often comes up and seems to be a topic everybody thinks they know something about. For the last thirty four years Dr. Etchells and his colleagues have been teaching a senior elective on the topic based on their industrial experience at several universities. It is a synthesis course bringing together a number of previous topics such as momentum transfer and mass transfer along with aspects of reaction engineering. Topics such as non-Newtonian rheology, multi-phase and turbulence are looked at in a new way. It is a design oriented course with open ended problems and because of the incomplete nature of the field it is constantly evolving. Not everything is understood about fluid mixing. Examples will be given.

(291b) An Academics View On Mixing Education At the University of Alberta

Suzanne Kresta, University of Alberta, Edmonton, AB, Canada

Biography

Suzanne Kresta (BSc, UNB (1986), MSc, Leeds (UK, 1987), PhD, McMaster (1992)) teaches Mass and Energy Balances (CME 265), design (ChE 464) and Mixing (ChE 420 and 620), and serves as co-editor of the Handbook of Industrial Mixing. Her principle research contributions are related to understanding turbulence and mixing in stirred tanks (measurement of turbulence, the use of spatial statistics to define mixing and length scales, nano-particle production, solids suspension, and reactor design) and she has collaborated with colleagues in a wide variety of industries (including oils sands extraction and froth treatment, photographic film, drinking water treatment, mineral processing and metals refining, polymer reactor design, cosmetics, mixing equipment design, fish processing, clean-up of nuclear waste, and fertilizer production). Many of her papers are both widely cited in the research literature and used for industrial design calculations.

Her teaching interests include visual problem solving tools, the use of story telling and case studies, active learning, and explicit use of cognitive levels with learning objectives. She has served as an Iron Ring Warden since 1999, as a Peer Consultant at the University of Alberta since 2004, and has won a number of national and international awards.

Research Interests

The mixing unit operation is far broader than the classical well mixed continuous stirred tank reactor, and is present in most chemical processes. Tightly designed mixing equipment is needed to meet the demanding process specifications of products ranging from Corianä countertops, through pharmaceutical and photographic crystals, cosmetics, fine chemicals, water treatment, and extraction of metals from ore.

Our group is currently exploring the theory of spatial statistics and how it can be applied to mixing problems to provide a rigorous definition of well mixed. The equations underlying the modelling of mixing sensitive reactions are being revisited in this context. Applications projects include the production of nanoscale metal powder, the distribution and suspension of multi-modal particle mixtures, drawdown of floating solids, and dissolution kinetics of partially soluble liquids.

Current Projects

- Continuous production of nanoscale metal precipitates; Shad Siddiqui, Alena Kukukova, Peter Unwin; Umicore, NSERC
- Drawdown of floating solids; Oscar Khazam, Lightnin
- Modelling equations for mixing sensitive reactions; Imran Shah, NSERC
- Segregation dimensions and the characterization of mixing; Suzanne Kresta, Alena Kukukova, Joelle Aubin; NSERC, CNRS
- Multimodal solids suspension and distribution; Inci Ayranci, Lightnin

(291c) Integrating Mixing Education Into Fluid Mechanics and Reaction Engineering Courses At Rowan

Robert P. Hesketh, Chemical Engineering, Rowan University, Glassboro, NJ

At Rowan University mixing is integrated in several courses:

Process Fluid Transport

Chemical Reaction Engineering

Unit Operations Experimental Design and Analysis

As well as taught in a senior elective titled, "**Biopharmacuetical and Industrial Fluid Mixing**"

Topics in Process Fluid Transport:

1 week is spent on Mixing:

Mixing Definition:

Mixing is the reduction of inhomogeneity in order to achieve a desired process result.

Inhomogeneity: concentration, phase, temperature.

Process results: increase mass and/or heat transfer, reaction rate, or product properties.

Typical Tank Dimensions

Types of Tanks with mixers

Fluid flow patterns with different impellers, diameters and liquid viscocity. PBT: Pitched Blade Turbine, FBT: Flat Bladed Turbine.

Impeller Types and Power Numbers

Power and D/T

Scale-up

Blend Time

Homework: Calculate the required blend time for the following tank:

Pitched blade turbine (6 Blade)

Impeller Diameter: 1 m

Operating speed: 84 RPM

Fluid: Water

Tank Diameter = 3.0 m

Tank Volume = 5600 gallons

1. Calculate Reynolds Number
2. Calculate blend time for within 5% of desired value
3. Design a blending tank for a volume of 10,000 gallons assuming =H and =/3 and dimensions scaled with volume
    1. Constant Blend Time
    2. Constant speed
    3. equal power per unit volume
    4. assuming power per unit volume based on Figure 15.

Give the power required, the tank and impeller dimensions and impeller speed for each case.

Topics in Chemical Reaction Engineering

Industrial Mixing in Chemical Reactors, Richard Grenville, DuPont

Biopharmacuetical and Industrial Fluid Mixing

The senior elective is taught by Drs. Arthur W. Etchells and Richard Grenville

In both the biotechnology and the pharmaceutical industries mixing is an important step in development, scale up and manufacture. It is most important in process steps involving multiple phases and complex liquids and chemical reactions. This course will discuss the application of fundamental single and multi-phase fluid dynamics to such processes and the development of industrial equipment designs based on these fundamentals along with the concepts of process scale up and scale down. A major objective of the class is to develop equipment designs based on fundamentals and empirically derived correlations. Emphasis will be on designs for the biotechnology and pharmaceutical industry but having equally applicability to other process industries. Both lecturers have extensive industrial experience with Dupont and other companies.

Arthur W. Etchells is a world recognized authority in the field of mixing for the process industries. For thirty nine years he worked for the DuPont Company and for thirty years as an internal consultant for the many diverse DuPont businesses in the field of fluid flow with emphasis on mixing and slurry transport and retired in 2002. He is a past president of the North American Mixing Forum and winner of their award for contribution to mixing technology. He is now an independent consultant in the field.

Richard K. Grenville is the lead consultant in mixing for the Dupont Company. He has twenty five years experience in the field. He has extensive publications in the field, has given many seminars to companies outside of Dupont and is a member of the North American Mixing Forum governing board.

(291d) Ready to Use Module for Introducing Mixing in a Fluids Mechanics Course

Richard K. Grenville, DuPont Engineering Technology, Wilmington, DE

As Chemical Engineers we are concerned with the transformation of raw materials into valuable products. These transformations may require the mixing and reacting of the raw materials and the separation of the desired products from by-products. Also, as these transformations take place, the materials may need to be heated and/or cooled. All the disciplines required to understand the operation and design of mass and heat transfer equipment are covered in the under-graduate Chemical Engineering curriculum.

Mixing has been identified as a critical operation for the Chemical Industry (see for example, Smith, 1990, *Trans IChemE.* 68, pp. 3 - 6) yet it is not, generally, taught to Chemical Engineering students. There are a number of reasons for this. Among those given are:

It is the University's job to "teach the basics" and the engineer will learn "practical skills" once he/she has started work.

In order to add a course on mixing, some other subject would have to be removed from the curriculum.

But the same comments are not made about heat exchangers or distillation columns.

If we accept that there is no "market" for a full mixing course at most Universities, how can we ensure that the students get some exposure to this technology before they graduate? We believe that a working understanding of mixing can be taught by adding lectures as mixing modules to the classic Chemical Engineering courses. For example:

An agitator impeller is a pump and it can be characterized in the same way as the pump in a piping system that students will have studied in Fluid Mechanics.

Design of agitators for heat and mass transfer can be covered in the traditional courses.

Mixing issues relating to fast, competitive reactions can be covered in the traditional Reaction Engineering course. When does the assumption that a reactor is "perfectly-mixed" fail?

The content of several proposed modules will be presented

(361a) Talking about Multi-Scale Simulations

Harry E.A. Van den Akker, Dept. of Multi-Scale Physics, Delft University of Technology, Delft, Netherlands

Many industrial multi-phase flows inherently are multi-scale systems in which several time and length scales interact in a complex way: particle (or bubble or drop) size, the spectrum of turbulent eddies, the scale of the equipment. Diffusion, heat conduction, and chemical reactions may complicate the scene even further at the higher end of the spectrum. And simulating these systems computationally adds another scale, viz. that of the computational cell size. Various options are available for such simulations, each with their own pros and cons. Most of the options heavily rely on such divergent issues as the concept of interpenetrating continua, particle size distribution kernels, sub-grid scale models, models for the interaction force(s) between the phases, models for turbulence under two-phase flow conditions, micro-mixing models, et cetera. For the simulation of industrially relevant systems, clever choices should be made, with due regard to the size of the computational effort affordable.
An essential question is: Does the set of particle differential equations really reflect the actual physics of the system of interest? Some dilemmas in answering this key question are discussed: how realistic is it to track many (point) particles in an Euler-Lagrangian approach? What are the limitations of the Euler-Euler method? How should we take sub-grid scale effects into account?
Current computational resources may make it possible to invoke the promise of Direct Numerical Simulations carried out in periodic boxes. DNSs refrain from models and are capable of resolving the real physics. Carrying out a number of such DNSs for conditions spanning the spatial distribution of governing parameters might reproduce the actual processes of interest much better. Most computational time should be devoted to mimicking the detailed processes, while much less efforts is spent in the coarse-grained reactor-scale simulation. Feeding back the results of the localized DNSs back into the coarse-grained simulation and interpolating between the results of the set of localized DNSs is part of the challenge of this truly multi-scale approach.

(361b) Multi-Scale Approach in Mixing for Colloidal Systems Design: Concentrated and Highly Concentrated Inverse/Direct Emulsions

Diego Pradilla[1], Andrea Nieto[1], Natalia Murcia[1] and Oscar Alberto Alvarez[2], (1)Chemical Engineering Department, Universidad de Los Andes, Bogotá, Colombia, (2)Chemical Engineering Department, Universidad De Los Andes, Bogotá, Colombia

Emulsions are found in many fields of science: they can be found in cosmetics, pharmaceuticals, oil extraction and food industry. An emulsion is a type of colloid in which there are two phases, normally one is dispersed as droplets in other one. This dispersion is thermodynamically and kinetically unstable, reason why an agent that reduces the interfacial tension is needed (i.e surfactant). Energy consumed, concentration, type of mixing system, droplet size distribution and stability, are some of the key variables in emulsion design.

During an emulsification process, several variables related to process and formulation must be taken into account because the final properties of the product depend on the condition at which each one of them is set. Relationships in such mixing systems are usually established by analyzing the effect of one variable on a property of the final product; however, this does not allow the observation of other interactions that might arise at different levels.

In this paper, the coupled effect between molecular interactions, the microscopic structure, the macroscopic level response and the energy consumed by direct oil-in-water and inverse water-in-oil emulsions was studied. A semi-batch emulsification process was used changing the type of impeller in the mixing system and the concentration of the dispersed phase. Near-infrared spectroscopy (NIR) was used to study the molecular response, laser diffraction (Mastersizer 3000) was used to obtain the particle size distribution, the rheological behavior was measured through steady state and oscillatory tests using a controlled stress rheometer (ARG2 TA Instruments) and the energy consumed was obtained through torque measurement from the mixing system (Lightnin LabMaster Mixer).

Using the multi-scale approach to analyze the final properties and characteristics of a product, different parameters can be evaluated simultaneously at the molecular, microscopic and macroscopic level and its relationship with emulsification variables.

For instance, the change in the microscopic structure of the water droplets influences a macroscopic response of the emulsion and a relationship between these two levels can be established through the pumping capacity of the mixing system.

(361c) Lattice Boltzmann Simulations of Drop Deformation and Breakup in Shear Flow

Alexandra E. Komrakova, Department of Chemical and Materials Engineering, University of Alberta , Edmonton, AB, Canada and Jos Derksen, Department of Chemical & Materials Engineering, University of Alberta, Edmonton, AB, Canada

Mixing of two immiscible liquids in tanks and agitated vessels is commonly encountered in chemical, petroleum, food, and pharmaceutical industries. When liquids are brought into contact and subjected to mechanical agitation, droplets formation occurs. The behavior of liquid-liquid dispersions in agitated systems is related to the flow conditions that vary from laminar to fully-developed turbulent flow. These hydrodynamic conditions together with interfacial tension between two liquids determine the breakage and coalescence rates of the drops. Drop breakup and coalescence are two fundamental processes that define the most important characteristic of a dispersion: the drop size distribution (DSD). Typically, in industrial applications, coalescence is suppressed by surfactants, and dispersions are subjected to turbulent flow conditions. In that case the dispersion formation and final DSD are mostly determined by the breakage process. Drops intensively break in the impeller region characterized by high shear stresses and as a result of local interactions with turbulent eddies.

In the present study the deformation and breakup of single organic drops in water subjected to simple shear flow were simulated using a diffuse interface free energy lattice Boltzmann method (LBM). Pure binary systems characterized by low surface tension, such as *n*-butanol/water, without mass transfer between phases were considered. Fully three-dimensional simulations of droplets between moving plates were performed. To validate and verify the developed numerical technique, simulations of several test cases were carried out. The results of test cases were compared to experimental data and results obtained with other numerical methods. A wide range of flow conditions determined by capillary and Reynolds numbers was examined that allowed to study the behavior of such systems in great detail. As an example of simulation results, the evolution in time of *n*-butanol drop shape in shear flow is shown in the figure below. The proposed numerical approach can be straightforwardly implemented to the study of dilute Newtonian emulsions in complex flow.

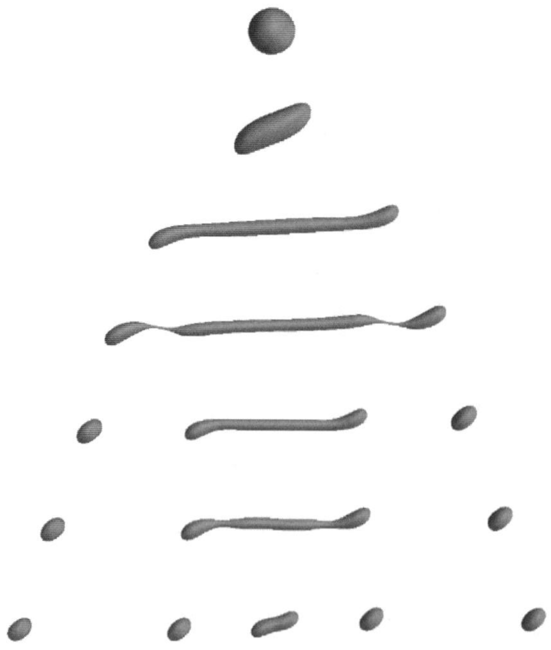

(361d) Nanostructure of Individual Grains Affecting Movement in Densely Packed Material

Casey Kick, Xan Voon and **Kalyana Pingali**, Paper Engineering, Chemical Engineering and Imaging, Western Michigan University, Kalamazoo, MI

Ultrathin nanolayers deposited on pharmaceutical particles during the mixing process were investigated with respect to their powder flow behavior. Multiple pharmaceutical formulations were exposed to various shear rate and shear strain conditions in a blender. Samples of lubricated and un-lubricated blends were collected from the blender after the mixing process and their micro and nanostructures were studied under field emission scanning electron microscope (FESEM) and transmission electron microscope (TEM). The powders with such nanostructures tested for their flow properties in a gravitational displacement rheometer (GDR) showed that the powder flow depended not only on the density of packing of powder bed but also on nanostructures of individual grains. Velocity profiles of powder particles in a tumbling cylinder were tracked at different rotational rates ranging from 1 – 20 rpm. It was interesting to see that the ultra-thin nanolayers affected the velocity distribution of particles within the powder bed as a function of dilation and rotational rate. An attempt made to track the magnitude of each particle average velocity from the interior 3D system is expected to open an innovative approach in particle dynamics.

(361e) A Multiscale Model for the Granulation of Powders in a Spheronizer

Jonathan Bouffard, Jamal Chaouki and **Francois Bertrand**, Department of Chemical Engineering, Ecole Polytechnique de Montreal, Montreal, QC, Canada

The spheronizer is a rotor-based equipment used in the pharmaceutical industry as a mixing and pelletizing process to obtain uniformly sized spherical pellets from a mixture of powder extrudates, the form of which is cylindrical. It consists of a rotating disk that spins at high speed at the bottom of a cylindrical bowl, thereby increasing friction with the cylindrical particles. Powder flow in a spheronizer is characterized by a toroidal motion that promotes the roundness of the pellets and increases their density. One important issue related to this process is the likely occurrence of segregation within the bed of particles, which can lead to a lack of uniformity of the pellets and, subsequently, to a poor quality of the end-use product. Granulation is a process whereby primary particles mix and adhere to one another to form larger entities called granules. A standard spheronizer can be easily transformed into a granulator by adding to it an atomizer that sprays water droplets to the surface of the bed of particles. Despite the fact that the mechanisms governing the formation of these granules are very complex, their understanding is crucial to the development of efficient granulation processes yielding granules with the desired properties. Granulation processes can be modeled with the help of population balance models. Along this line, numerous studies have considered mechanisms such as coalescence or breakage to predict the evolution of granulation with respect to time. Although these mechanisms play a fundamental role in representing the real behavior of such systems, it has become evident that particle flow also needs to be modeled. This presentation will focus on the development of a multiscale model involving a population balance model at the process scale, a compartmental approach that takes into account specific flow behaviors in different zones of the equipment, and a discrete element based model at the particle scale. After introducing briefly this model, simulation results will be presented and compared to experimental data obtained in our pilot-scale rig.

(361f) Solids Suspension Simulations At Multiple Scales

Jos Derksen, Department of Chemical & Materials Engineering, University of Alberta, Edmonton, AB, Canada

Solids suspension simulations at multiple scales

Solids suspension in mixing tanks is a multi-scale process with the tank size as the macro-scale and the particle size (or the interparticle spacing) as micro-scale. Resolving all scales in a single simulation is not doable given the usually large width of the length scale spectrum. Simulation approaches to the suspension process recognize this and parameterize part of the length-scales range: In Euler-Euler approaches individual particles are not considered and the focus is on the global (turbulent) flow characteristics and the distribution of the phases in the tank. Euler-Lagrange simulations do consider individual solid particles but not the details of the flow around them. Simulations that do resolve the flow around particles ('particle-resolved simulations') are feasible but have severe limits in terms of the number of particles and the sizes of the systems to be simulated (certainly much smaller sizes than even a lab-scale mixing tank). In this paper we suggest strategies how to relate simulations at different levels to enhance our understanding of the solids suspension process. The emphasis will be on what happens near the bottom of a tank in a not completely suspended situation. Here particles forming a granular bed on the bottom get entrained by the flow generated by the impeller. As an example, the figure shows a particle-resolved simulation of the start-up of this entrainment process in a miniature mixing tank.

Figure. Start-up of the suspension process for monosized spherical particles. Realizations after (from left to right) 2, 4 and 32 impeller revolutions.

(404a) Mixing Issues in Stem Cell Culture Bioreactors Using Microcarriers

Alvin W. Nienow[1], Qasim A. Rafiq[2], Karen Coopman[2] and Chris J. Hewitt[3], (1)Birmingham University, Birmingham, UK, (2)Chemical Engineering, Loughborough University, Loughborough, United Kingdom, (3)Chemical Engineering, Loughborough University, Leicestershire, United Kingdom

Though in the early 1980s animal cell culture was generally based on attaching cells to microcarriers, little work was done to optimise their suspension. Partly this was because, around this time, free suspension culture was established where cells were much less likely to be damaged by fluid dynamic stresses. For example, Croughan et al (1987) showed that mean specific energy dissipation rates had to be < ~$1x10^{-3}$ W kg^{-1} to prevent damage during cell culture with 180 mm microcarriers whilst values up to 0.25 W/kg are possible in free suspension (Nienow, 2006). However, the optimisation of vessel/agitator configurations to achieve low mean specific energy dissipation rates suspension $(W/kg)_S$ for were not investigated and some geometric recommendations (for example, the use of hemispherical bioreactor bases to aid suspension (van Wezel, 1985)) were counter to the general findings on particle suspension (Nienow, 1985, 1997). Now with the increasing importance of regenerative medicine and the need to grow stem cells on microcarriers at large scale for allogeneic usage, establishing efficient bioreactor geometries for microcarrier suspension has become important. Clearly, microcarriers suspension is essential if the well-mixed features of the stirred bioreactor are to be achieved. The minimum speed, N_{JS}, and mean specific energy dissipation rate at that speed, $(W/kg)_S$, has been measured for many geometries but still little work has been done on microcarriers. The most relevant study (Ibrahim and Nienow, 2004) investigated N_{JS} using Cytodex 3 microcarriers with different diameter Chemineer HE-3 hydrofoils, a pair of Ekato InterMIG impellers and a six-blade, 45°-pitch turbine impeller in a baffled vessel of 19.2 L operating volume containing phosphate buffer saline solution. Flat and modified tank bases were used and N_{JS} values were observed to be in the range of 50 to 90 rev min^{-1}. The use of Zwietering's correlation with geometric suspension parameters, S, from the earlier literature would have predict N_{JS} values up to 50% higher. The low N_{JS} values obtained were attributed to the very small particle–liquid density difference (40 kg m^{-3}), which eased the lifting of the particles from the tank bottom, compared to those used in non-microcarrier studies. With these microcarriers, the three-blade hydrofoil HE-3 impeller of D/T = 0.39 in a cone-and-fillet based tank was marginally the most efficient; that is, it had the lowest $(W/kg)_S$ of ~ 0.5 x 10^{-3} W kg^{-1}. However, $(W/kg)_S$ < ~1 x 10^{-3} W kg^{-1} in most cases with the different size HE-3 hydrofoils, which implies that these geometries should be suitable for shear-sensitive microcarrier cell culture systems. When cells are attached to microcarriers, then damage might occur to cells from impacts between impeller and microcarrier, microcarrier and microcarrier and microcarrier and vessel internals (as in secondary nucleation (Nienow and Paul, in press)); or due to cells being stripped off them and direct stress from turbulent eddies. Thus, it is important to minimise $(W/kg)_S$. However, it must also be remembered that the mean specific energy dissipation rate is a critical parameter and needs to be sufficient to achieve the required rate of oxygen and CO_2 transfer and give an adequate homogenisation. Thus, it is important to consider all parameters which are dependent on it when developing the stirred bioreactor. This approach has led to the successful culture of mesenchymal stem cells in a 5L bioreactor.

(404b) Multi-Phase Drop Size Distribution Determination in a Centrifugal Contactor

Nicholas B. Wyatt[1], Timothy J. O'Hern[2] and Bion Shelden[1], (1)Sandia National Laboratories, Albuquerque, NM, (2)Engineering Sciences, Sandia National Laboratories, Albuquerque, NM

Centrifugal contactors are widely used in separation and mass transfer applications involving immiscible phases. Quantifying drop size distributions in such processes is critical to understanding and tuning mass transfer and separation characteristics between immiscible phases. Many studies have focused on the complex flows in both the annular and separation regions of the contactor. However, the magnitude and distribution of droplet sizes in the mixing region has not been characterized. For mass transfer applications, drop size and, thus, surface area are key factors. Through the use of high speed imaging and laser fluorescence, we measure drop size distributions in the annular mixing region for mixtures of silicone oil and water. We measure the evolution of the drop size distribution as the material travels from the mixing region to the vane region. Further, we quantify size distributions for air bubbles that are entrained in the flow. The distributions are examined over a range of rotor speeds from 1100 to 3200 rpm for silicone oil viscosities from 5 to 20 cSt. Drop sizes for both air and oil (dispersed phase) are found to be normally distributed at locations near the surface of the mixing zone. Near the bottom of the mixing zone, the distributions become log-normal and the mean drop diameter shifts to lower values.

Sandia National Laboratories is a multi-program laboratory operated by Sandia Corporation, a wholly owned subsidiary of Lockheed Martin Corporation, for the U.S. Department of Energy's National Nuclear Security Administration under contract DE-AC04-94AL85000.

(404c) Development and Application of a CFD Model for Gas-Droplet Multiphase Flow with Evaporation in Optimizing Spray Injection and Atomization for Control of Reactor Selectivity

Hua Bai, The Dow Chemical Company, Freeport, TX, Michael D. Cloeter, Core R&D, Engineering Sciences, Fluid Mechanics & Mixing, The Dow Chemical Company, Freeport, TX and Subrata Sen, Engineering Sciences, Core R&D, The Dow Chemical Company, Freeport, TX

This presentation describes the development and application of a comprehensive computational fluid dynamics (CFD) model pertaining to a spray technology. The spray nozzle was proposed for recycle of a liquid stream into the feed pipe of a gas-phase reactor. The reactor produces two chemical products sold into two different markets. The recycle injection impacts reactor selectivity, thus improves the process capability in responding to transient market conditions. Process reliability requires a complete evaporation of liquid droplets within the feeding pipe prior to entering the reactor with only a fraction of a second residence time available. In addition, contact between the droplets and the feed pipe wall needs to be minimized in order to prevent a liquid film wall layer leading to potential fouling due to thermal polymerization.

The developed CFD model integrated comprehensive physics - including turbulent hydrodynamics, multiphase flow, convective heat transfer, phase change (droplet evaporation), and latent heat effect - into an Euler-Lagrange framework with 2-way coupling for gas/droplet interactions. The motion of droplets is affected by the continuous gas phase convection, and as well the droplets and their evaporations affect the flow of the gas phase. The droplet size distribution used in the model was based on experimental measurements. The dispersion of droplets due to turbulence was predicted using the stochastic tracking (Random Walk) model that includes the effect of instantaneous turbulent velocity fluctuations on the droplet trajectories during integration through the use of stochastic methods. However, achieving a fully converged solution for such a complicated model was time-consuming. Therefore, different simplification strategies were adopted in different phases to accelerate the design process by reducing the turnaround time for CFD simulations. The developed spray technology was implemented in the plant and has shown the desired improvement in product flexibility.

(404d) Continuous Powder Mixing Monitoring Using Multiple Multivariate Models with NIR Spectroscopy Online

Luis Obregon, Chemical Engineering, University of Puerto Rico Mayaguez, Mayaguez, PR and Leonel Quiñones, Chemical Engineering, University of Puerto Rico - Mayaguez Campus, Mayaguez, PR

NIR spectroscopic was used to get on-line spectra and obtain immediate results of API concentration at the output of a continuous powder mixer in order to assess the efficiency of the mixing. The results were compared with values obtained with ultraviolet spectrophotometer. Multiple Calibration curves were developed prior to continuous mixing experimentation with UV and NIR methods with blends ranging in API concentration from 0% to 100% (w/w). DmodX and Hotelling were the statistical functions used to identify the appropriate model online with an implementation of a control logic to perform the prediction with the best exactitude and precision with values around 2-5% and 1-3% w/w respectively depending on the value of concentration studied. The calibration models were generated taking into consideration the variability of the process working at different RPMs of the mixer and at different feed rates of the feeders and considering as well the powders properties mainly the powder cohesion variation in the mentioned range of API concentration. During the process of mixing three different step changes were applied to the feeders to obtain three different API concentrations in the range 0-30% to characterize the dynamic response of the continuous mixer. The results obtained by UV and NIR showed stable mixing at the different API concentrations and achieved a reasonable mixing with values of RSD lower than 6%. The error obtained with the NIR and UV increases when increasing the API concentration and it was attained to the increase in powder cohesion which made more difficult the process of mixing. The parameters of the dynamic response were similar to those obtained theoretically.

(404e) Dispersion in a Resonating Acoustic Mixer

Samin Rezania[1], Jonathan Michael D. Thomas[2], Amlan Chakraborty[2] and **R. Eric Berson**[2], (1)University of Louisville, Louisville, KY, (2)Chemical Engineering, University of Louisville, Louisville, KY

We previously showed that overcoming mass transfer limitations by using an impeller-less resonating acoustic mixer during enzymatic saccharification of pretreated corn stover slurries maintained glucose release rates and yields despite orders of magnitude increases in viscosity as solids concentrations increased up to 30% w/w. In this study, dispersion coefficients in the mixing vessel were determined as a function of viscosity in order to quantify the mass transfer improvements and determine if the system was a valid choice for studying reactions in the absence of mass transfer limitations that are typical of highly viscous systems. Since conventional means of measuring dispersion could not be performed due to the slurry characteristics, a computational fluid dynamics model was developed that simulated motion of the slurries in the mixer. The model was very well validated by comparing experimentally determined mixing times using an electrolytic tracer with mixing times from the simulations. Computational and experimental mixing times differed by 2% or less in all cases. Axial dispersion coefficients were generally 3 to 6 times higher than radial dispersion coefficients. Despite the high viscosities studied here, up to 25,000 cP, dispersion values were achieved simlar to water flow in a pipe, indicating the system was a valid choice for studying reactions in the absence of mass transfer limitations.

APPLICATION OF ECCENTRICALLY AGITATED STIRRED TANKS IN BIOTECHNOLOGY SETTINGS: EXAMPLES, ADVANTAGES, AND CHALLENGES

Mario M. Alvarez[1], Pamela Belén Sánchez-Arreola[1], Josefina Castillo-Reyna[2], David Bulnes-Abundis[1], Marisa Granados-Pastor[1], María del Refugio Rocha-Pizaña[1], and Leydi Maribel Carrillo-Cocom[1].

(1) Centro de Biotecnología-FEMSA, Tecnológico de Monterrey at Monterrey, Monterrey, México;
(2) Departamento de Ingeniería Química y Biotecnología, Tecnológico de Monterrey, Monterrey, México.

ABSTRACT: In this contribution, we provide examples of the use of eccentrically agitated stirred tanks in biotechnology applications and discuss their potential advantages. At the lab scale and in industrial practice, stirred tank bioreactors are the most common cell culture platform. However, these bioreactors are designed based on conventional stirred tanks geometries (namely, tanks agitated with concentrically located Rushton turbines or axial impellers), which are not necessarily ideal for some cell culture applications. We used tanks agitated with angled (45°), eccentrically located disc impellers to explore four different cell culture scenarios: (a) culture of a recombinant strain of *Escherichia coli* (a producer of a recombinant influenza antigen); (b) culture of Chinese Hamster Ovary (CHO) cells; (c) small scale culture of human hematopoietic stem cells, and (d) culture of tobacco (*Nicotiana tabacum*) plant cells. The recombinant *E. coli* cultures showed experimentally calculated specific growth rates and maximum cell densities in 30mL eccentric bioreactors that were significantly lower than those obtained in fully instrumented 1L bioreactors, and the recombinant product yield was at least 30% higher. The CHO cell cultures showed similar specific growth rates and the maximum cell densities in the small scale eccentric bioreactor to those determined in culture bottles (30 mL), Erlenmeyer flasks (125 mL), or fully instrumented bioreactors (1L). Comparatively, eccentrically stirred tank systems provide good mixing at low RPM values (and low shear rates). Consequently, these systems could be advantageous for the culture of presumably shear sensitive organisms, such as plant or mammalian cells. Human hematopoietic stem cell populations were successfully expanded 10-fold in suspension conditions using a 30mL eccentric mini-bioreactor system. Lastly, tobacco cells were sustained in good condition (viability above 80%) for longer times when cultured in instrumented 1L eccentric bioreactors compared to conventional concentrically stirred tanks.

In all these case models, the cell culture performance was similar (if not better) than that observed in conventional culture systems. Our results suggest that eccentrically stirred tank geometry provides superior mixing performance, practicality, and scalability.

KEYWORDS: CHO cells, eccentric agitation, cell culture, bioreactor, stirred tank

* Ave. Eugenio Garza Sada 2501 Sur Col. Tecnológico C.P. 64849, Monterrey, Nuevo León, México, +52 (81) 8358-2000, +52 (81) 8328-4136, mario.alvarez@itesm.mx.

INTRODUCTION

Stirred tanks are the most commonly used bioreactor system in pilot plant and industrial settings. In this contribution, we present examples of the use of tank vessels that are eccentrically agitated with angled discs in cell culture settings. Four different scenarios are presented: (a) culture of a recombinant strain of the aerobic bacterium *Escherichia coli* (a producer of a vaccine candidate); (c) culture of Chinese Hamster Ovary (CHO) cells (producers of a follow-on monoclonal antibody (mAb)); (d) small scale culture of human hematopoietic stem cells, and (d) culture of tobacco (*Nicotiana tabacum*) plant cells. The study of eccentrically agitated systems as mixing vessels has recently received research attention (Alvarez et al., 2002; Ascanio et al., 2002; Karcz et al., 2005; Montante et al., 2006; Sánchez-Cervantes et al., 2006; Szoplik and Karcz, 2008; Cabaret et al., 2008; Yang et al., 2008; Woziwodzki and Jedrzejczak. 2010; Hu et al., 2010; Yang et al., 2011; Takahashi et al., 2011). However, the use of eccentricity in the context of biotechnology applications has not been documented. In this work, we extend the discussion of potential, advantages, and challenges of eccentrically stirred tanks to the area of cell culture applications.

MATERIALS AND METHODS

Stirred tank systems

Vessels 30 mL and 1L in volume were eccentrically agitated using an angled (90°) disc impeller. In these systems, the degree of eccentricity (E) was defined as the ratio of the distance between the shaft and the vertical centerline of the tank (*e*) and the tank radius (R): $E = e/R$.

Cell culture experiments in the eccentric 30 mL bioreactor

Cell culture experiments were performed in 30 mL eccentric reactors (E=0.42) agitated by an angled impeller (Figure 1). Three different cell models were tested: (a) a Rosetta-gami™ strain of *E. coli*, from Novagen® (EMD Cat. Num. 71350) engineered to produce an influenza vaccine antigen; (b) a CHO-S cell line; a producer of a monoclonal antibody, and (c) human hematopoietic stem cells (CD34+) isolated from the umbilical cord blood.

The recombinant *E. coli* recombinant strain was cultured in LB media, according to standard protocols. Temperature was maintained at 37 °C. No pH control was implemented. Samples were taken to monitor biomass concentration (by absorbance at 600 nm). The specific growth rate was calculated by assuming a first-order kinetic growth model; namely, $d[X]/dt = \mu[X]$ (where [X] is the biomass concentration at any given time, and t is time). The specific growth rate (μ) during the exponential phase can be calculated from the slope of the linearized and integrated biomass material balance: $\ln X = \mu t + \ln [X]_o$.

A series of CHO cell culture experiments (three replicates) were conducted using a CD OptiCHO™ culture medium (Gibco ®, Invitrogen Cat. Num.12681-029). A cell concentration of $2X10^5$ cells/mL was used as inoculum, the culture temperature was maintained at 37 °C, and the culture medium was changed on the fourth day. Culture samples were taken every 24 hours. Cell viability was estimated by staining cells with trypan blue and counting using a Neubauer chamber. In these cultures, the specific growth rate during the exponential phase (μ [=] h^{-1}) and the maximum cell concentration, $[X]_{max}$, were calculated and compared against those observed in agitated six-well culture plates (V=2 mL), Erlenmeyer flasks (V=125 mL), and fully instrumented bioreactors (1L) DASGIP® (Germany).

Human hematopoietic stem cells (CD34+) isolated from umbilical cord blood were cultured in DMEM F-12 (Invitrogen Cat. Num. 12400-024 Carlsbad, USA) supplemented with 10% FBS (Invitrogen Cat. Num. 16000-044, Carlsbad, USA), 1% penicillin-streptomycin solution (Sigma, Cat. Num. P-4333, St. Louis, USA), and 10 mM HEPES (Gibco, Invitrogen Cat. Num.11344-041, Carlsbad, USA). Cells were seeded at a concentration of $1.75X10^4$ cells/mL. The culture medium (10% FBS, 1% penicillin-streptomycin solution, 10 mM HEPES) was changed every 24 hours, according to a fed batch protocol. Briefly, agitation was suspended for five minutes to allow cells to settle to the bottom of the vessel, and then 80% of the culture medium was removed from the upper portion of the tank using a sterile syringe and replaced with fresh medium at 37 °C. Samples of 1 mL cell suspension were removed for counting every 36 hours. Cells were stained with 10% Trypan blue and counted using a Neubauer camera.

Cell culture experiments in other cell culture devices

For comparison, additional *E. coli* cell culture experiments were conducted in fully instrumented 1L bioreactors (Applikon, Netherlands) with pH and dissolved oxygen control. The agitation speed, temperature, and dissolved oxygen in these bioreactors were set at 500 RPM, 37 °C, and 20%, respectively. The specific growth rate (μ[=] h^{-1}) during the exponential phase and the maximum cell concentration $[X]_{max}$ were calculated in all of these systems and compared against those observed in the 30mL eccentric bioreactor.

Additional CHO cell culture experiments were carried out in: (a) six-well culture plates (with no pH or dissolved oxygen control); (b) 125mL Erlenmeyer flasks (with no pH or dissolved oxygen control); and (c) fully instrumented 1L bioreactors (DASGIP, Germany). The agitation speed, temperature, and dissolved oxygen in the bioreactors were set at 150 RPM, 37 °C, and 30%, respectively.

Culture of tobacco cells

Tobacco (*Nicotiana tabacum* 1;NT-1) cells were cultured in different bioreactor types including 1 L eccentrically stirred tank systems. Tobacco cell suspensions were cultured in NT1 medium containing 4.3 g Murashige and Skoog basal salts, 30 g sucrose, 50 mL MES (20x stock), 10 mL B1-Inositol (100x stock), 3 mL Millers I (stock) and 2.2 mL 2,4-D (stock) in deionized water. The pH of the medium was adjusted to 5.7 with KOH. Culture experiments were conducted in 1 L Erlenmeyer flasks containing 500 mL NT1 medium.

Flasks were inoculated by adding 5 g/L of callus previously established on agar plates and incubated in an orbital shaker at 150 rpm and a temperature from 25 to 27°C. The medium was changed weekly. Flasks were sampled for analytical measurements by removing a 5 mL volume daily. Specific growth rates at the exponential phase, maximum cell densities, and mortality rates were calculated for each flask experiment. In addition, culture experiments were conducted in 350 mL instrumented concentrically stirred tank bioreactors (Infors®, Switzerland) stirred at 100, 200, and 300 RPM. Initial 5 g/L cell concentrations were used. Additional culture experiments were performed in a 2L instrumented bioreactor (Applikon®, Netherlands). The reactor stainless steel multiport head plate was modified to allow for concentric and eccentric agitation. Temperature was set at 26°±1°C in all bioreactor experiments. Cell biomass was reported as dry cell weight, determined from well mixed samples taken at different culture times. Due to the small sample size, the samples were freeze dried. Viability determinations were made daily by staining with fluorescein diacetate (FDA) and propidium iodide (PI). Cells were counted in triplicate in different microscope fields. The viability data were used to estimate mortality rates for the different culture conditions explored.

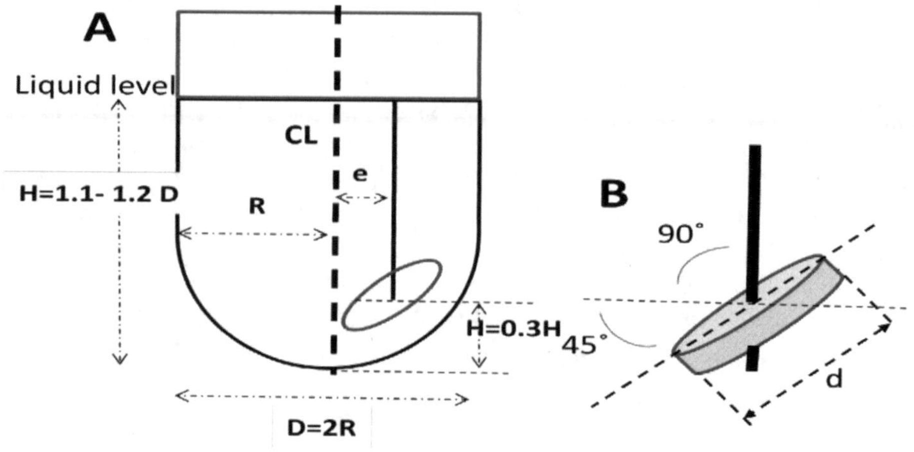

Figure 1. Geometrical features of stirred tanks agitated by eccentrically located angled disks. (a) Tank geometry; (b) details of the disc impeller.

RESULTS AND DISCUSSION

Cell culture of recombinant aerobic bacteria in eccentric bioreactors

E. coli is the most commonly used bacteria for the recombinant expression of proteins. *E. coli,* like other general aerobic bacterial cultures, demands high amounts of oxygen for adequate growth. The transfer of oxygen through the direct injection of air is a routine strategy in lab, pilot plant, and industrial instrumented bioreactors. However, in a small scale stirred tank, direct air injection would cause flooding and a loss of volume. We therefore investigated if oxygen mass transfer rates from the liquid surfaces would be sufficient to support low density aerobic bacterial cell growth and recombinant protein production in our 30 mL eccentric bioreactor by following the evolution of *E. coli* cultures, measuring the optical density over time and calculating the specific growth rate (μ [=] h^{-1}]) and maximum cell density ($[X]_{max}$ [=] absorbance units) at different RPM values.

Figure 2a shows the concentration of biomass and recombinant protein in an *E. coli* culture experiment that used an agitation rate of 500 RPM. No oxygen was supplied by bubbling; therefore, all oxygen transfer occurred through the gas-liquid interface. Protein production was induced by injection of IPTG at time=7 h. We observed adequate production of recombinant protein ($[P]_{max}$=3.8 g/L), at levels 30% higher than those observed in fully instrumented bioreactors (see Figure 2a and 2c). However, the values of μ for this recombinant strain of *E. coli* were lower than those observed in 1L fully instrumented bioreactors. The specific growth rate calculated during exponential growth phase was μ=0.475 h^{-1} (Figure 2b) and 0.69 h^{-1} in eccentric mini-reactors and fully instrumented 1L bioreactors, respectively.

The maximum biomass concentration is a good indirect indicator of oxygen mass transfer efficiency in miniature aerobic systems. In our experiments, $[X]_{max}$ was 50% higher in the eccentric mini-bioreactor agitated at 500 RPM than at 400 RPM. Similarly, [P]max was 1.66 times higher in systems agitated at 500 RPM. Practically the same $[X]_{max}$ was observed in eccentric systems agitated at 500 RPM and in fully instrumented bioreactors, but the time to achieve this value was much shorter in instrumented systems. These results suggest that oxygen limitations in the eccentric mini-bioreactor still allow for the production of high concentration of protein.

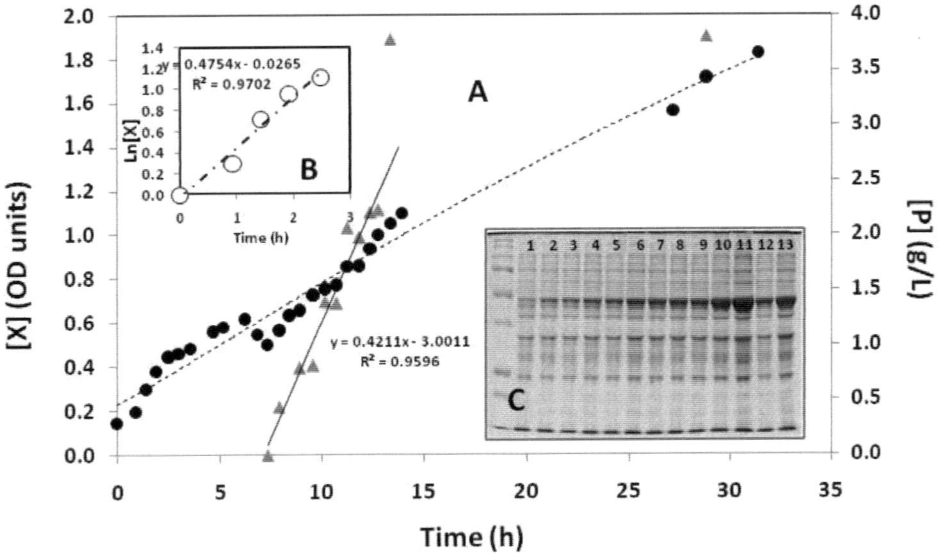

Figure2. Production of a recombinant protein by *E. coli* cultured in a 30 mL eccentrically agitated stirred tank. **A.** Concentration of biomass (•) and recombinant protein (▲) in time. **B.** Plot of Ln([X] vs. time to calculate the initial specific growth rate of the culture (μ=0.475 h^{-1}). C. SDS-PAGE showing the accumulation of the recombinant product over time. Lanes 1 to 13 correspond to samples taken from the reactor at different times after induction, in the range from 1 to 30 h.

Culture of mammalian cells in eccentric bioreactors

CHO cells are the most commonly used mammalian host for the commercial production of therapeutic glycosylated proteins (Coco-Martin and Harmsen, 2008; González-Leal et al., 2011, Kim et al., 2012). They are used, in particular, for the commercial production of monoclonal antibodies. Figure 3a shows the cell count profiles for experiments where 2.5X10^5 CHO cells were seeded into a 30 mL eccentric bioreactor operated at 250 RPM with no pH control. No air was bubbled into the tank; oxygen mass transfer occurred exclusively through the gas/liquid interface. The exponential growth phases of the three growth cultures overlapped. Although the extent of the lag phases varied between these experiments (not shown), remarkable consistency was observed during exponential growth. Cell densities higher than 1X10^6 were achieved after 96 hours of cultivation. This evolution and cell density was also similar to those observed during the same time period in fully instrumented bioreactors (see Figure 3a). Specific growth rates and $[X]_{max}$ values calculated from experiments in four different commonly used lab scale culture systems are graphically compared in Figure 3b.

Statistically similar μ values were observed in all systems; however, the $[X]_{max}$ values were not statistically equivalent among all culture systems. Again, $[X]_{max}$ can be considered as an indirect indicator of oxygen transfer efficiency in small-scale systems. Fully instrumented bioreactors and eccentric mini-bioreactors were statistically equivalent in terms of $[X]_{max}$. Erlenmeyer flasks gave higher $[X]_{max}$ values, while 6-well microplates showed the statistically lowest $[X]_{max}$ values (Figure 3b). In summary, considering both parameters (μ and $[X]_{max}$), 30mL eccentrically agitated mini-bioreactors performed equivalently to 1L fully instrumented bioreactors for the culture of CHO cells. The $[X]_{max}$ values measured in 30 mL and 1L bioreactors were also similar to those observed in full commercial-scale operations – on the order of 5 X10^6 cells/mL (Nienow, 2006).

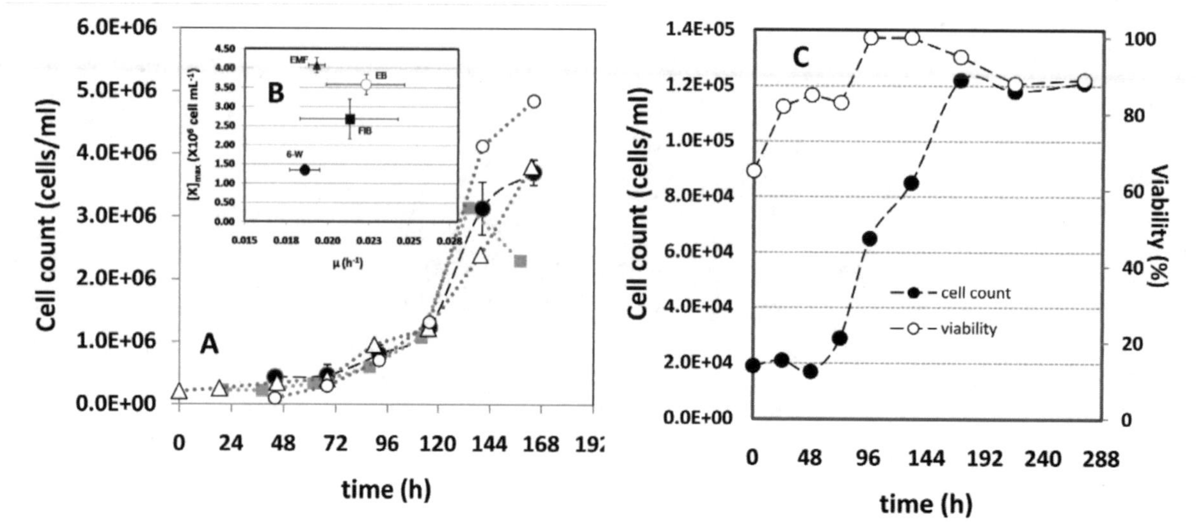

Figure 3. Culture of mammalian cells in different bioreactors: **A.** Average cell concentration profile in time observed during the batch culture of a CHO cell line in a 30mL eccentrically stirred tank agitated by an angled disc (•) and in three independent culture experiments in 1L instrumented concentric bioreactors (Δ;■,○); **B.** Comparison of specific growth rate (m) and maximum cell concentration $[X]_{max}$ for CHO cell cultures in different lab-scale bioreactors: a 6 well culture microplate (6-W), Erlenmeyer flasks (EF); fully instrumented 1L bioreactors (FIB), and 30 mL eccentrically agitated bioreactors (EB). C. Cell counts (•) and the viability (○) during an experiment of expansion of hematopoietic cells derived from umbilical cord conducted in a 30 mL eccentrically agitated bioreactors.

Proliferation of hematopoietic stem cells in eccentric bioreactors

Massive stem cell expansion has been recognized as a bottleneck for the development of feasible tissue engineering treatments. The culture of stem cells in stirred tank bioreactors has only recently been explored (Rodrigues et al., 2011; Kowalczyk et al., 2011; Want et al., 2012). In our experiments with hematopoietic stem cells, agitation was set at 200 RPM to allow for adequate mixing while minimizing damage induced by mechanical stress. Although adequate liquid-liquid mixing is achievable at much lower speeds, visualization experiments with neutrally buoyant 250 micron particles (results not presented here) demonstrate that higher speeds are needed to hold cells in suspension. The just-suspended velocity for polystyrene particles in water was estimated at 200 RPM. Cells cultured in the eccentrically stirred tank bioreactor at an initial concentration of $1.75X10^4$ cells/mL exhibited growth curves (Figure 3c) with characteristic latency, exponential, and stationary phases. Cell doubling time, calculated using a linear regression analysis on the exponential growth phase, was 31.46 hours, which was a 19.18% increase from doubling time in stationary culture Petri dishes (standard culture protocol for this type of cell; Murugappan et al., 2010).

Cell viability was highly maintained, in the range of 80% to 100% (Figure 3c). In the exponential phase, cell viability values were sustained in the range of 95 to 100%, which was direct evidence that the flow field conditions are acceptable for sustaining growth of these extremely shear-sensitive cells. The lack of blades is partially responsible for lower stresses at impeller discharge. In addition, the impeller angle creates an oscillating motion in the impeller vicinity, helping to disperse the high shear effect over a wider area. Cells remained non-differentiated (CD 34^{+}) during culture, as demonstrated by expression analysis using PCR protocols (not shown).

Growth of tobacco cells in eccentrically agitated bioreactors

Additional experiments were conducted at different inoculum concentrations and using different operational strategies in stirred tanks agitated with inclined disc impellers mounted on eccentric shafts. Eccentric bioreactor configurations have been confirmed as adequate for proper mixing at low Reynolds regimes, thereby avoiding cellular damage induced by intense agitation (Alvarez et al., 2002; Sánchez-Cervantes et al., 2006).

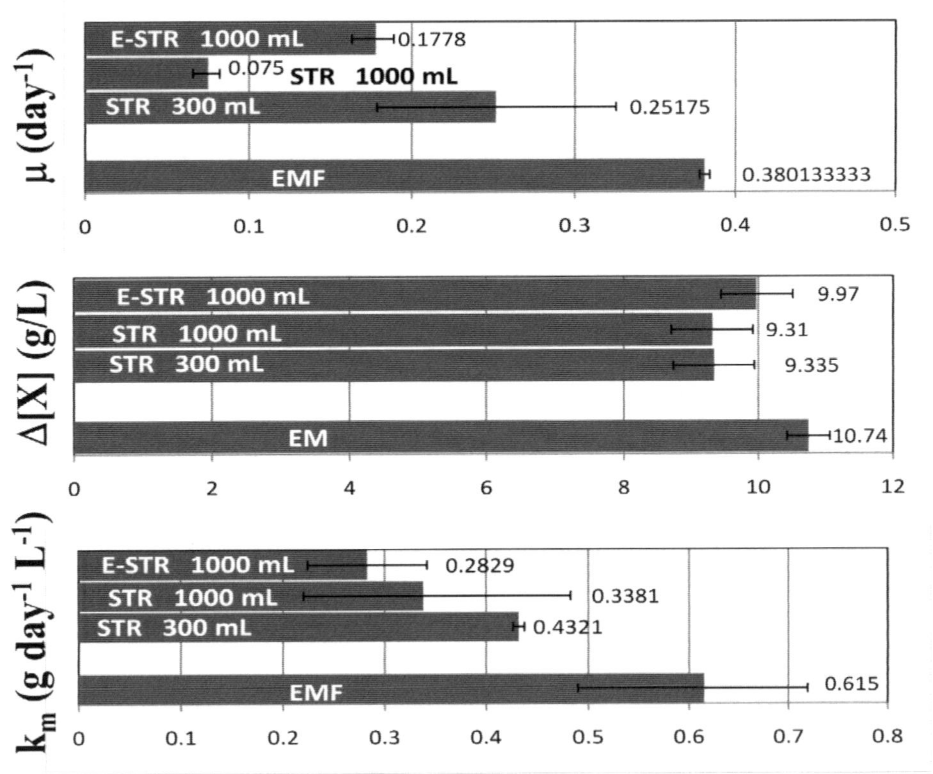

Figure 3. Comparison of the specific growth rate (μ), the maximum cell density [X_{max}], and the mortality rate (k_m) of tobacco cells cultured in different bioreactor systems: 1L eccentrically stirred tank bioreactors (E-STR 1000mL), 1L concentrically stirred tanks (STR-1000mL); 300mL concentrically stirred tanks (STR-1000mL); and Erlenmeyer Flasks (EMF).

Adequate cell densities and high growth rates were expected for these culture experiments. In experiments at a low inoculum concentration of 1 g/L, lower specific growth rates were observed than occurred in Erlenmeyer flasks. The best results were obtained with high inoculum concentrations (5-6 g/L), which gave specific growth rates similar to those observed in Erlenmeyer flasks. Two stage protocols in eccentrically agitated bioreactors were also tested. In high inoculum experiments, cells were cultured under eccentric agitation at low speed (50 RPM) for several days (during exponential growth) and then agitation speed was increased to 300 RPM. Based on the scaling-up criteria of maximum tip speed velocity, this agitation condition would be comparable to more than 360 RPM in the 350 mL concentric bioreactors described previously.

Nevertheless, no cell damage or increased mortality rate were observed in high density cultures in eccentric bioreactors. Indeed, viabilities above 80% were sustained for twelve days of culture (Figure 4). Similar or better values of relevant biological indicators (μ, $[X]_{max}$, k_m) were observed in high inoculum eccentrically agitated bioreactors when compared with Erlenmeyer flasks or concentrically agitated stirred tanks of similar scale. Our experimental results suggest that eccentrically agitated systems can be used as a cell culture system for shear sensitive plant cells.

CONCLUSIONS

This paper reports cell culture applications for stirred tanks agitated with an angled disc impeller mounted in an off-centered shaft. The system was used for low volume applications related to: (a) the production of a recombinant influenza antigen in an engineered *E. coli* strain; (b) the study of growth kinetics of a CHO cell line; and (c) the expansion of hematopoietic stem cells derived from umbilical cordon. In addition, culture experiments using tobacco NT cells were conducted in 1L fully instrumented eccentric bioreactors using two-stage culture protocols.

In all of these cases, eccentric systems performed equivalent to, or better than, conventional culture systems in terms of critical output process variables such as specific growth rates and maximum biomass values. In the particular case of aerobic bacterial cultures, issues related to oxygen mass transfer still limit the application of the proposed system to low scale and low to medium cell density applications.

ACKNOWLEDGEMENT

The authors acknowledge the financial support provided by the Mexican government through CONACYT scholarships for PBSA, JCR, DBA, and MGP. The authors are grateful for the financial support provided by Tecnológico de Monterrey at Monterrey through Cátedra de Investigación (CAT 122).

REFERENCES

1. Alvarez M.M., Arratia P.E., and Muzzio F.J. (2002). Laminar mixing in eccentric stirred tank systems. *Canadian Journal of Chemical Engineering, 80*(4): 546-557.
2. Ascanio G., Brito-Bazán M., Brito-De La Fuente E., Carreau P.J., and Tanguy P.A. (2002). Unconventional Configuration Studies to Improve Mixing Times in Stirred Tanks. *The Canadian Journal of Chemical Engineering,* 80(4): 558-565.
3. Cabaret F., Fradette L., and Tanguy P.A. (2008). Effect of shaft eccentricity on the laminar mixing performance of a radial impeller. *The Canadian Journal of Chemical Engineering,* 86(6): 971-977.
4. Coco-Martin J. M., Harmsen M. M.: A review of therapeutic protein expression by mammalian cells. *Bioprocess International*, 4:28-33, 2008.
5. González-Leal I.J., Carrillo-Cocom L.M., Ramírez-Medrano A., López-Pacheco F., Bulnes-Abundis D., Webb-Vargas Y., and Alvarez M.M. (2011). Use of a plackett-burman statistical design to determine the effect of selected amino acids on monoclonal antibody production in CHO cells. *Biotechnology Progress,* 27 (6): 1709-1717.
6. Hu Y., Liu Z., Yang J., and Cheng Y. (2010). Liquid Mixing in eccentric stirred tank. *Huagong Xuebao/CIESC Journal*, 61 (10): 2517-2522.
7. Karcz J., Cudak M., and Szoplik J. (2005). Stirring of a liquid in a stirred tank with an eccentrically located impeller. *Chemical Engineering Science,* 60(8-9): 2369-2380.

8. Kim J.Y., Kim Y-G., and Lee G. (2012). CHO cells in biotechnology for production of recombinant proteins: current state and further potential. *Applied Microbiology and Biotechnology.* 93(3): 917-930.
9. Kowalczyk M., Waldron K., Kresnowati P., and Danquah M.K. (2011). Process Challenges relating to hematopoietic stem cell cultivation in bioreactors. *Journal of Industrial Microbiology and Biotechnology,* 38(7): 761-767.
10. Montante G., Bakker A., Paglianti A., and Magelli F. (2006). Effect of the shaft eccentricity on the hydrodynamics of unbaffled stirred tanks. *Chemical Engineering Science*, 61 (9): 2807-2814.
11. Murugappan G., Carrillo-Cocom L.M., Johnson K.E., Bulnes-Abundis D., Zavala-Arcos J., Moreno-Cuevas J.E., González-Garza M.T., and Alvarez M.M. (2010). Human Hematopoietic Progenitor cells grow faster under rotational laminar flows. *Biotechnology Progress,* 26(5): 1465-1473.
12. Nienow A.W. (2006). Reactor engineering in large scale animal cell culture. *Cytotechnology,* 50(1-3): 9-33.
13. Rodrigues C.A.V., Fernandes TG, Diogo M.M., da Silva C.L., and Cabral J.M.S. (2011). Stem cell cultivation in bioreactors. *Biotechnology Advances,* 29(6): 815:829.
14. Sánchez-Cervantes M.I., Lacombe J., Muzzio F.J., and Alvarez M.M. (2006). Novel bioreactor design for the culture of suspended mammalian cells. Part 1: Mixing characterization. *Chemical Engineering Science*, 61(24): 8075-8084.
15. Szoplik J. and Karcz J. (2008). Mixing time of a non-Newtonian liquid in an unbaffled agitated vessel with an eccentric propeller. *Chemical Papers,* 62(1): 70-77.
16. Takahashi K., Shigihara D., and Takahata Y. (2011). Laminar mixing in eccentric stirred tank with different bottom. *Journal of Chemical Engineering of Japan*, 44 (12): 931-935.
17. Want A.J., Nienow A.W., Hewitt C.J., and Coopman K. (2012). Large-scale expansion and exploitation of pluripotent stem cells for regenerative medicine purposes: beyond the T flask. *Regenerative Medicine,* 7(1): 71-84.
18. Woziwodzki S. and Jedrzejczak, Ł. (2010). Effect of eccentricity on laminar mixing in vessel stirred by double turbine impellers. *Chemical Engineering Research and Design,* 89(11): 2268-2278.
19. Yang F.L., Zhou S.J., and Wang G.C. (2011). Experimental study and detached eddy simulation of the macro-instability in an eccentric stirred tank. *Applied Mechanics and Materials*, 66-68: 20-26.
20. Yang, F.L., Zhou, S.J., Zhang, C.X., and Chen, L.F. (2008). Investigation on solid-liquid suspension performance in an eccentrically stirred tank. *Guochengg Gongcheng Xuebao/The Chinese Journal of Process Engineering,* 8(6): 1064-1069.

(526a) The Mixing Mechanism Change and Age Distribution in a Kenics Static Mixer

Minye Liu, DuPont Engineering Technology, Wilmington, DE

Scalar mixing and Age distribution in a Kenics static mixer will be studied by species concentration and mean age solutions. In order to resolve diffusion effect, a microscaled mixer is used by CFD method. The spatial distribution of mean age reveals locations of older material that could potentially cause process fouling and product defects. At the micro scales of the mixer, both convection and diffusion strongly affect concentration and age distributions. A method is developed to compute an average effective diffusivity to measure false diffusion due to numerical errors. The range of Peclet number is found that the scalar solution is not affected by false diffusion. The effect of convection and diffusion is studied for Re numbers that cover the full range of laminar flows in the mixer. It is found that a mixing mechanism change results a transition of mixing efficiency. Diffusion broadening is also found to have strong effect on the striation structures in the mixer. Mixing and age distribution in the mixer is quantitatively measured with frequency distribution function and two different variances, the variance of residence time and the variance of mean age. It is found that concentration variance reaches a peak while mean age variance decay rate increases almost monotonically as Re increases.

(526b) Analysis of Melt Flow of Bridgman Process Using Lagrangian Coherent Structures

Mojtaba Izadi, Chemical & Materials Engineering, University of Alberta, Edmonton, AB, Canada, Youssef Belhamadia, University of Alberta, Edmonton, AB, Canada and Stevan Dubljevic, Chemical and Materials Engineering, University of Alberta, Edmonton, AB, Canada

Bridgman process is one of the commonly used methods for the production of melt-grown crystals. In the inverted Bridgman method, the lower zone of the furnace with temperatures above the melting point of the material is separated from the upper zone with a temperature below melting point by an adiabatic baffle. The crucible containing the solid material is lowered into the hot zone and after temperature stabilization, the crucible is raised slowly into the upper zone, so that the crystal grows from the melt.

In this process, thermal convection plays an important role by affecting heat and mass transfer. It has been demonstrated that during the formation of a crystal from its melt, mixing properties of the melt flow affect transport phenomena. If complete mixing in the melt does not occur, a gradient of impurities concentration exists in the melt near the crystal interface.

In the framework of dynamical systems theory, it is well-known that there exist special flow patterns in a fluid domain which are the skeletons of observed tracer patterns and govern transport structure and mechanics of the flow. These patterns are referred to as coherent structures and when captured in terms of quantities derived from particle trajectories, they are called Lagrangian coherent structures (LCS). Recently, effective computational algorithms have been developed for the identification of these structures based on the eigenvalues and eigenvectors of the finite-time deformation tensor.

In this work, we study the kinematics of mixing of the melt flow in Bridgman process by computing the LCS in the melt domain. To this end, velocity field is obtained based on the mixed finite element resolution of the coupled governing equations including Navier-Stokes, continuity and energy equations. Having velocity data, fluid element trajectories are obtained numerically and used for the computation of deformation tensor at each point in the melt domain, which finally reveals the LCS. Providing a novel insight into the transport of the flow, LCS can be considered as stable and unstable manifolds in the phase space of the dynamical systems. Because all of these manifolds are invariant, meaning that fluid particles do not cross them, the fluid that is trapped in the intersection of these manifolds (lobes) is confined to remain in them as time evolves. Therefore, the motion of lobes in terms of entrainment and detrainment helps explain the transport and mixing processes.

References:

1. G. Haller, Distinguished Material Surfaces and Coherent Structures in Three-Dimensional Fluid Flows, Physica D 149, 2001, 248--277.

2. S. C. Shadden, F. Lekien, J. E. Marsden, Definition and Properties of Lagrangian Coherent Structures from Finite-Time Lyapunov Exponents in Two-Dimensional Aperiodic Flows, Physica D 212, 2005, 271--304.

3. S. C. Shadden, J. O. Dabiri, J. E. Marsden, Lagrangian Analysis of Fluid Transport in Empirical Vortex Ring flows, Phys. Fluids 18, 2006, 047105.

4. G. Haller, A Variational Theory of Hyperbolic Lagrangian Coherent Structures, Physica D 240 (7), 2011, 574--598.

(526c) Effect of Impeller Submergence, Position, Diameter and Speed On Flow Regime, Surface Air Entrainment and Mixing Effectiveness of Partially Filled Stirred Vessels

Shilan Motamedvaziri and **Piero Armenante**, Otto H. York Department of Chemical, Biological and Pharmaceutical Engineering, New Jersey Institute of Technology, Newark, NJ

In many industrial applications, mixing vessels have a liquid height-to-tank diameter ratio, H/T, equal to, or larger than 1. However, there are many instances where this ratio is lower than 1, as in all those cases in which the vessel is either emptied or filled. Even when $H/T<1$, sufficient agitation must still be provided in order to attain the desired process objectives. When the impeller submergence is reduced as a result of lowering the liquid level, the fluid dynamics of even a single-phase stirred liquid can become quite complex, with different regimes possibly existing depending on the geometric characteristics of the system (such as impeller clearance, liquid height, or liquid submergence above the impeller). The objectives of this work were to study in detail the hydrodynamic changes that occur when H/T is decreased, and to determine the minimum liquid levels and the critical impeller submergence for different impeller off-bottom distances, impeller diameters and agitation speeds where adequate mixing process can still be achieved, both in a single liquid phase and in solid-liquid suspensions. A number of experimental tools were used to analyze the systems under investigation, including: Particle Image Velocimetry (PIV) for the experimental determination of the velocity profiles, and after the appropriate data manipulation, the impeller pumping capacity and its Pumping Number; a strain gage-based rotary torque transducer system to measure the power dissipated by the impeller; and a visual observation method to determine the minimum agitation speed for complete solids suspension, N_{js}. In addition, Computational Fluid Dynamics (CFD) modeling was used to predict the behavior of the system in terms of its velocity profile, Power Number, and Pumping Number. CFD predictions were obtained using a multiple reference frame (MRF) model coupled, when needed, with a Volume of Fluid (VOF) model, in order to study systems in which a vortex could be expected to form.

In general, good agreements between the experimental data and the predicted results for the velocities distribution, Power Number, radial Pumping Number, mixing time, and N_{js} were obtained. Results show that there is a critical impeller submergence ratio S_b/D below which: (1) the macroscopic flow pattern generated by the impeller changes substantially, transitioning from either a “double-loop” recirculation flow or a "single-loop" recirculation flow (depending on the impeller clearance off the tank bottom) to an upward "single-loop" recirculation flow; (2) the Power Number and radial Pumping Number drop significantly; (3) solid suspension cannot be attained at any agitation speed; (5) vortex formation occurs, air entrainment is significantly facilitated, and impeller flooding typically results. This is the first time that such hydrodynamic regime change has been reported and characterized. When this flow regime transition occurred, it was observed that the average velocity field and turbulence intensity close to the tank bottom decreased substantially; this was identified as the reason why solid suspension became unachievable. Furthermore, the critical impeller submergence ratio resulting in the establishment of the newly described flow regime was not affected by dynamic variables such as agitation speed. Impeller flooding was observed only when the new regime was established and when the vortex depth reached the impeller disk. This phenomenon was correlated to critical values of the Froude number. Additionally, by decreasing the impeller off-bottom clearances the operating window within which the stirred vessels could be operated effectively became wider. However, for low impeller off-bottom clearances ($C_b/T<0.05$) and low impeller submergence ratios ($S_b/D<0.37$, corresponding to $H/T<0.16$) even the lower recirculation loop was suppressed (tank bottom effect) and an unstable flow system was observed with no clear recirculation, implying that effective mixing could not achieved.

(526d) Stretching Properties of the Flow and Impinging Jets Stability Studies in RIM

Cláudio P. Fonte, M. Ashar Sultan, Ricardo J. Santos, Madalena M. Dias and José Carlos B. Lopes, LSRE - Laboratory of Separation and Reaction Engineering - Faculty of Engineering - University of Porto, Porto, Portugal

Reaction Injection Molding (RIM) is an industrial process for the production of plastic parts where two or more viscous liquid monomers are injected and mixed in a cylindrical mixing chamber by opposed impinging jets. The mixture that leaves the mixing chamber is discharged into a mold where the polymerization occurs and the plastic part is formed. The mixing chamber is a cylinder with 1 cm diameter and 5 cm length, the injectors' diameters are in the range of 1 – 2 mm and the mean residence time in the mixing chamber is in the range of 10 – 100 ms. An efficient mixing of the two monomers in the mixing chamber before entering the mold must be achieved in order to avoid zones of unreacted monomers that may lead to structural defects and the eventual rejection of the plastic parts [1]. The jets' Reynolds number range of industrial practice goes up to 600. Previous computational and experimental works [2, 3] have shown that mixing in RIM mainly occurs above a critical jets' Reynolds number that marks the transition from a steady segregated laminar flow regime to a self-sustained laminar chaotic regime with strong mixing dynamics. It has also been shown that in equal injector diameter chambers the iso-momentum condition between the opposed jets is fundamental to ensure the continuous presence of the self-sustainable chaotic flow regime [4].

In this work the stretching properties of the flow in the RIM machine mixing chamber are studied from 3D Computational Fluid Dynamics (CFD) simulations and Lagrangian tracking of the position and deformation of material spherical elements. The flow regimes and the jets' impingement point position under unequal flow rates and different injectors' diameters are studied from PIV experiments. An elastic analogue model of the jets is proposed to predict its impingement point.

Results show that the flow under certain conditions develops the ability to reorient trajectories and produce a nearly exponential average stretching of fluid elements with time, a feature of chaotic flows. Both CFD simulations and experimental results show that $Re > 110$ is the value that sets the transition from a steady laminar flow regime to a self-sustained laminar chaotic flow regime. The good agreement between the PIV experiments and the elastic analogue model indicates that the jets' kinetic energy feeding rate ratio is the key parameter to control mixing in RIM for formulations with stoichiometric and density ratios different from 1.

[1] P. Kolodziej, W.P. Yang, C.W. Macosko, S.T. Wellinghoff, Impingement Mixing and its Effect on the Microstuture of RIM Polyurethanes, Journal of Polymer Science, Part B: Polymer Physics, 24 (1986) 2359-2377.

[2] R.J. Santos, A.M. Teixeira, J.C.B. Lopes, Study of mixing and chemical reaction in RIM, Chemical Engineering Science, 60 (2005) 2381-2398.

[3] R.J. Santos, E. Erkoç, M.M. Dias, A.M. Teixeira, J.C.B. Lopes, Hydrodynamics of the mixing chamber in RIM: PIV flow-field characterization, AIChE Journal, 54 (2008) 1153-1163.

[4] M.I. Nunes, R.J. Santos, M.M. Dias, J.C.B. Lopes, Micromixing assessment of confined impinging jet mixers used in RIM, Chemical Engineering Science, 74 (2012) 276-286.

(526e) Investigating On the Influence of Mesh Density, Turbulence Models and Numerical Aspects of CFD Modeling in PBT Impellers Using Non-Structured Meshes

José Roberto Nunhez, Chemical Process Department, UNICAMP, Campinas, Brazil, Suzanne Kresta, University of Alberta, Edmonton, AB, Canada, Nicolas Spogis, ESSS, Sao Paulo, Brazil, Marcio Bezerra Machado, Chemical and Materials Engineering, University of Alberta, Edmonton, AB, Canada and José L. Gomez, School of Chemical Engineering, State University of Campinas, Campinas, Brazil

The use of non-structured meshes has received recently considerable attention for discretization of the computational domain in CFD simulations (Vakili and Esfahany, 2009; Cabrales, et al., 2011). The non-structured meshes (the most common are tetraedrals) allow adequate adaptation of irregular geometries and easy use of automated algorithms for mesh generation in most commercial CFD software. It has been consistently suggested in the literature that hexaedrical meshes are preferred over tetraedrals. However, they can be used, provided that some care is taken to provide a good representation of the physical phenomena of the problem. Recently published works have been proving that if they are used with care, satisfactory results are obtained (Spogis, 2007). Most of works developed with the use of CFD simulation for mixing processes use hexaedrals and to date tetraedrals do not have the same acceptance of hexaedrals (Joaquim, et al., 2007) and part of this work is aimed at showing that they can provide good results when simulating mixing processes.
Another objective of this work is to show the effect of mesh refinement in tetrahedrical meshes. It is investigated the influence of the prismatic elements near wall on global parameters such as the Power and Flow numbers. An initial analysis of the sensitivity of some turbulence models and the discretization scheme on the flow field produced were also carried out in this work. In addition, simulations with different mesh resolutions, height and number of elements near the wall were made using different turbulence models (Standard k-ε model, the shear stress transport (SST) and RNG k-ε model) and using the multiple reference frame (MFR) approach. The discretization schemes were limited to upwind and blend factor, since other schemes do not work well with tetraedrals. The commercial CFD software CFX 13.0 was used to simulate the results. A pitched blade turbine impeller (PBT 45° down-pumping), was simulated in turbulent flow.
The results obtained for the Power and Flow numbers, the axial velocity profile component of the model in CFD are discussed and compared with experimental data. The velocity was measured using the PIV technique. (Particle Image velocimetry data). A significant influence of some parameters is observed on the determination of some important mixing . In general, the predictions of the velocity profile are influenced in some way by the resolution of the mesh, turbulence model and the discretization.scheme.

(526f) Simulations of Mobilization of Bingham Layers in Agitated Tanks

Jos Derksen, Department of Chemical & Materials Engineering, University of Alberta, Edmonton, AB, Canada

Simulations of mobilization of Bingham layers in agitated tanks

Numerical simulations were used to study mobilization of a bottom layer of Bingham liquid by agitating a Newtonian liquid above the Bingham layer. The agitation is done by a pitched-blade impeller at a Reynolds number of 6000. The Bingham liquid and the Newtonian liquid are miscible. The parameter space of the simulations has a Bingham number Bn (defined as the ratio of yield stress over inertial stress) and a Richardson number Ri as dimensionless variables. The Richardson number quantifies the role of the density difference between the two liquids. The simulation procedure is based on the lattice-Boltzmann method for the flow dynamics, and a finite volume scheme to solve the local and time dependent composition of the liquid mixture. The moderate Reynolds number allows us to directly simulate the flow, without the use of turbulence closure or subgrid-scale models.

The simulations show how the bottom layer gets eroded by the turbulent flow above it (see the figure), and show that there exists a critical value of the Bingham number beyond which the turbulent flow is not able to remove the Bingham layer (compare the left and right panel of the figure). The Richardson number is less important for the mobilization process, but has impact on the time it takes for homogenization of the liquids after mobilization has occurred.

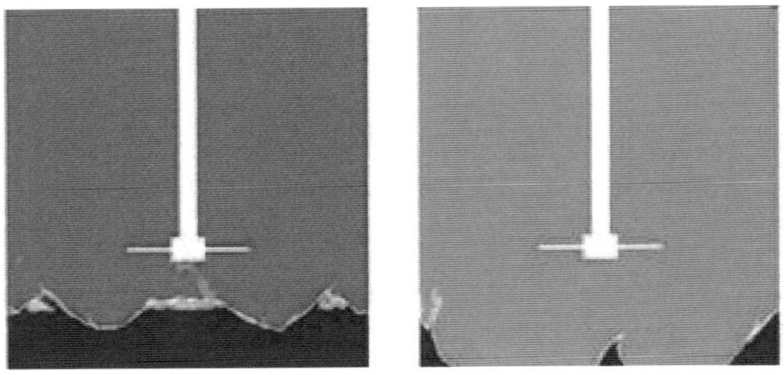

Figure: cross section through the mixing tank 100 impeller revolutions after start-up; left Bn=0.8, right Bn=0.1; blue indicates the Bingham liquid.

(625a) A Novel Euler-Lagrange Simulation Method for Gas-Liquid Stirred Reactors

Radompon Sungkorn, Chemical and Materials Engineering, University of Alberta, Edmonton, AB, Canada, Jos Derksen, Department of Chemical & Materials Engineering, University of Alberta, Edmonton, AB, Canada and Johannes G. Khinast, Institute for Process and Particle Engineering, Graz University of Technology, Graz, Austria

Various processes in chemical and biochemical industries involve dispersion of gas into liquid in stirred tank reactors. The complexity within the reactors are extremely high due to turbulent flow induced by impeller, interactions between the dispersed-gas and liquid phase, and interactions within the dispersed-gas phase. A simulation method that provides realistic description of the multiphase flow within the reactors is a valuable tool for design, optimization, and scale-up of such processes.

We present a novel simulation method for gas-liquid stirred reactors including bubble breakage and coalescence [1]. The filtered conservation equations for the liquid phase are discretized using a lattice-Boltzmann scheme. A Lagrangian approach with a bubble parcel concept is used to represent the dispersed-gas phase. Bubble breakage and coalescence are modeled as stochastic events. The momentum coupling between phases is realized through the source term in the conservation equations resulting in the so-called four-way coupling. The action of the reactor's components on the dispersed-gas and liquid phases is described using an immersed boundary condition.

The present method was used to simulate gas-liquid flow in a stirred reactor following the experiments of Montante et al. [2]. The predicted number-based mean diameter and long-term averaged liquid velocity components agreed qualitatively well with the experimental data (Fig. 1). Effects of the presence of the dispersed-gas phase and the gas flow rate on the hydrodynamics were numerically studied.

Furthermore, the present method was extended by including a power law model to represent shear-thinning liquids and empirical correlations for bubbles in shear-thinning liquids [3]. Simulations of aerated stirred reactors with shear-thinning liquids following the experiments of Venneker et al. [4] were carried out (Fig. 2). The predicted flow field of a single-phase stirred reactor with shear-thinning liquid showed reasonable agreement with the experimental data. For gas-liquid reactors, the predicted gas holdup distribution agreed qualitatively with the experimental data. Using the present method, it was found that a change in rheology significantly alters the number mean diameter, Sauter diameter, and the shape of bubble size distribution.

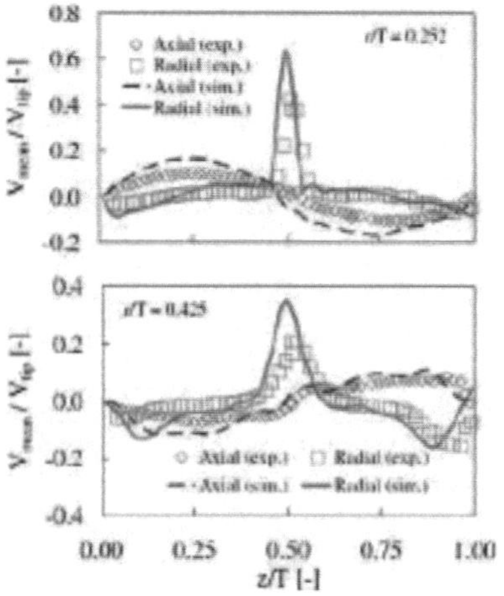

Figure 1. Comparison between experimental and predicted long-term average axial and radial liquid velocity.

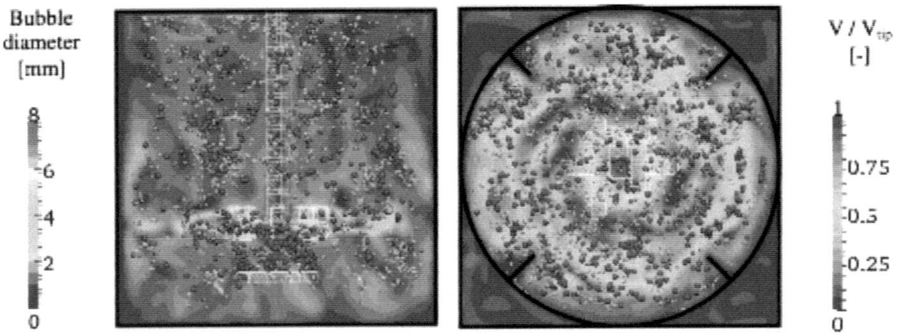

Figure 2. Snapshot of predicted bubble dispersion pattern in an aerated stirred reactor with shear-thinning power law liquid.

References

[1] Sungkorn R., Derksen J.J., Khinast J.G. Euler-Lagrange modeling of a gas-liquid stirred reactor with consideration of bubble breakage and coalescence. AIChE J. 2012;58:1356-1370.

[2] Montante G., Paglianti A., Magelli F. Experimental analysis and computational modelling of gas-liquid stirred vessels. Trans IChemE. 2007;85:647-653.

[3] Sungkorn R., Derksen J.J., Khinast J.G. Modeling of aerated stirred tanks with shear-thinning power law liquids. Heat and Fluid Flows. 2012 (accepted).

[4] Venneker V.C.H., Derksen J.J., Van den Akker H.E.A. Population balance modeling of aerated stirred vessels based on CFD. AIChE J. 2002;48:673-685.

Influence of the wall structure on the heat transfer in packed beds with small tube to particle diameter ratio

T. Eppinger (thomas.eppinger@tu-berlin.de), N. Jurtz, M. Kraume
Fachgebiet Verfahrenstechnik, TU Berlin, Straße des 17. Juni 136, 10623 Berlin, Germany;
N. Zobel, F. Behrendt
Fachgebiet Energieverfahrenstechnik und Umwandlungstechniken regenerativer Energien, TU Berlin, Fasanenstr. 89, 10623 Berlin, Germany

Aim of this contribution

The aim of this contribution is to show, firstly, that it is possible to obtain more homogeneous radial void fraction distribution in monodisperse packed beds of spherical particles by imposing appropriate wall structures and secondly, that a more homogeneous void fraction distribution leads to enhanced radial mixing and as a result to an improved heat transport in radial direction.

Introduction

Packed bed reactors are widely used in the chemical and process industry amongst others for catalytic surface reactions. The designs of such reactors are often done based on simplifying assumptions such as plug flow and a homogeneous void fraction distribution over the whole packed bed which are not valid for reactors with a small tube to particle diameter ratio. Such reactors are dominated by wall effects caused by the radial void fraction distribution. These wall effects constrain the heat transfer to or from the reaction zone which could lead to bad reactor performance, catalyst sintering or safety issues. Furthermore a reliable scale-up from lab-scale to full-scale apparatuses could be critical.

The effective thermal conductivity in radial direction can be considered as the superposition of the effective thermal conductivity of the solid particles which is a function of the local void fraction and the heat dispersion due to radial mixing of the fluid. This is illustrated by the empirical correlation for the effective radial thermal conductivity Λ_eff by Winterberg et al. (2000):

$$\Lambda_{\mathrm{eff}}(z) = \lambda_{\mathrm{bed}}(z) + \tfrac{1}{8}\varrho_f c_{p,f} u_{0,c}\, d_p f(z) \qquad (1)$$

with $f(z) = 1$ if $z > K\, d_p$

$$f(z) = \left(\frac{z}{K\, d_p}\right)^2 \quad \text{if } z < K\, d_p$$

and

$$K = 0.44 + 4\exp\left(-\frac{\mathrm{Re}}{70}\right)$$

with z being the distance from the wall, λ_{bed} the effective thermal conductivity of the bed with a stagnant fluid, ρ_f the density of the fluid, $u_{0,c}$ the superficial velocity of the fluid in the core region, d_p the mean particle diameter, $c_{p,f}$ the heat capacity of the fluid and f(z) a correction factor.

The extent of lateral mixing (second term in Eq. (1)) strongly depends on the extend of tortuosity and, thus, the local void fraction. If there is no particle (void fraction equal unity) disturbing the flow in axial direction, then both tortuosity and lateral mixing are minimal. The more particles are disturbing the flow in axial direction (i.e.: the lower the void fraction), the higher the tortuosity and, hence, the better the lateral mixing. Therefore it can be assumed, that if the radial void fraction can be homogenized with an appropriate wall structure, the heat transfer will be increased.

To explore the question how wall structures influence the fixed bed characteristics, Computational Fluid Dynamics (CFD) is used: Most of the results in this contribution are derived from numerical simulations and are validated against experimental results.

First, the Discrete Element Method (DEM) is used for the generation of a randomly packed bed. This method was first presented by Cundall et al. (1979). Spherical particles are randomly initialized in the reactor domain and fall due to gravity to the bottom of the reactor. For each particle a force balance is formulated and is solved for each time step which takes the interaction between particles and the interaction between particles and the surrounding reactor wall into account. Such a DEM simulation is finished when all particles are in rest. The results of such numerically generated beds have been presented in Zobel et al. (2012) and are in very good agreement with experimental data.

For the CFD simulation the 3D structure of the fixed bed has to be subdivided into small control volumes. While for the pure fluid dynamics calculation only the space between the particles has to be meshed, the particles have to be meshed for the heat transport simulations, too. It is well known from literature that the mesh quality in the vicinity of particle-particle or particle-wall contact points is crucial: Either the cells are highly skewed, which could lead to convergence problems or the mesh is highly refined in this region which increases the number of cells and the calculation time significantly. To elude this problem several methods are known, which are depicted in Figure 1: Some authors (Bai et al. (2009), Nijemeisland and Dixon (2001)) shrink the spheres by a certain extent to avoid contact points(Fig. 1-a) while others increase the size of the particles (Guardo et al. (2004)). Both methods modify the void fraction distribution which has to be corrected in subsequent step. Another method first described by Ookawara et al. (2007) (Fig. 1-b) is to bridge particles with small cylinders which are in contact with each other.

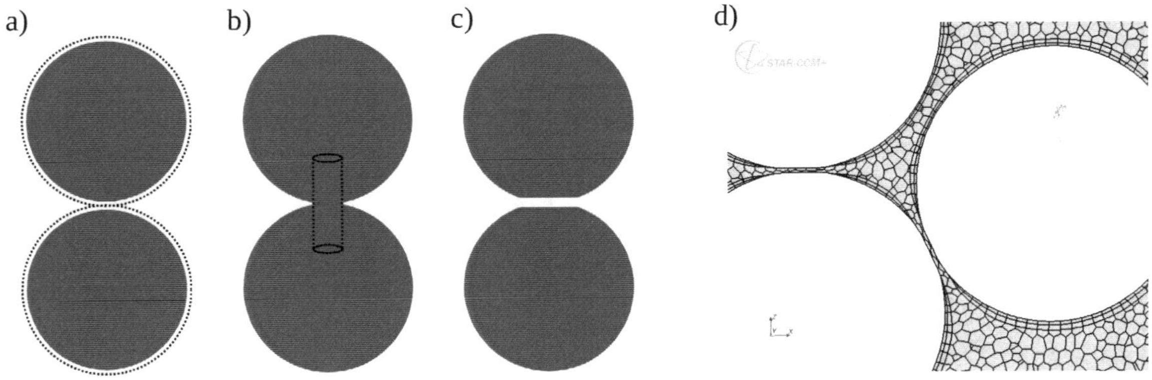

***Figure 1**: different ways to treat the meshing problems at contact points: a) reduction of the sphere diameter, b) bridging the contact point with cylinders, c) local flattening of the spheres and d) mesh example based on method c).*

In this work we use a recently developed method which flattens the particles at their contact points locally if a certain distance is underrun (Fig. 1-c). In this newly formed gap a small number of cells with good quality can be generated (Fig. 1-d). This method influences void fraction and therefore the pressure drop only slightly and – as the main advantage – the meshing can be done in a completely automated way. In Eppinger et. al (2011) it has been shown that with this method the radial, axial and global void fraction as well as pressure drop and radial velocity profiles agree quite well with literature data.

Methods

For the numerical simulations the commercial CFD code STAR-CCM+ version 6.06 by CD-adapco was used, which is based on the finite volume method and includes a DEM solver. The fixed bed consists of monodisperse glass spheres with a diameter of d=2.5 cm. The parameters used for the DEM calculation are listed in Table 1 and are typical values for glass spheres with a relatively rough surface. The generated bed is assumed to be in rest if all particles have a velocity of v < 1e-5 m/s. This bed is then used for meshing and for the CFD calculation. For meshing the base size is set to bs = 2.5 cm, which is equal to the sphere diameter. The triangle edge size of the sphere surface shells is 4-8 % of the basesize, while the minimum gap between two surfaces in the vicinity of the contact points is 12 % of the triangle edge size. Further settings are described in Eppinger et al (2011). The resulting mesh consists of 6-7 million polyhedron cells.

***Table 1**: DEM parameter*

Property	Symbol	Unit	Value
Diameter	d_p	m	0.025
Density	ρ	kg/m^3	2500
Coefficient of restitution, normal	e_N	-	0.9
Coefficient of restitution, tangential	e_T	-	0.5
Poisson coefficient	υ	-	235
Young's modulus	E	Pa	78.5E9
Sliding friction coefficient	μ_{sl}	-	0.2
Rolling friction coefficient	μ_r	-	0.002

The reactor model is depicted in Figure 2. The tube to particle diameter ratio is $D/d_p = 8.8$. For the CFD calculation gas flow with the properties of air (η= 1.577 e−5 m2 /s) is assumed. At the inlet a constant velocity profiles with v = 0.1 m/s is set ($Re_p = v*d_p/\eta = 158$). For the investigation of the radial velocity distribution constant density (ρ_f = 1.1767 kg/m3) and isothermal flow is assumed, while for the investigation of the heat transfer ideal gas is assumed and the inlet temperature is set to T_{inlet} = 298 K.

Six different wall structures and one reactor without wall structures have been investigated. These structures are depicted in figure 2 and consist of spherical caps derived from spheres with 2.5 cm diameter. Reactor R7 is the standard reactor without any wall structure, R4-R6 are reactors with wall structures with a regular arrangement of the caps. The wall structure and the fixed bed particle arrangement of reactor R1 is derived from the inner zone of a fixed bed with a larger D/d-ratio where the radial porosity distribution already reached a constant value. This inner zone has been cut out by a cylinder. This means that all particles lying completely inside the cylindrical cut were kept and particles, which were cut by the cylinder wall were used as wall structure. Reactor R2 is based

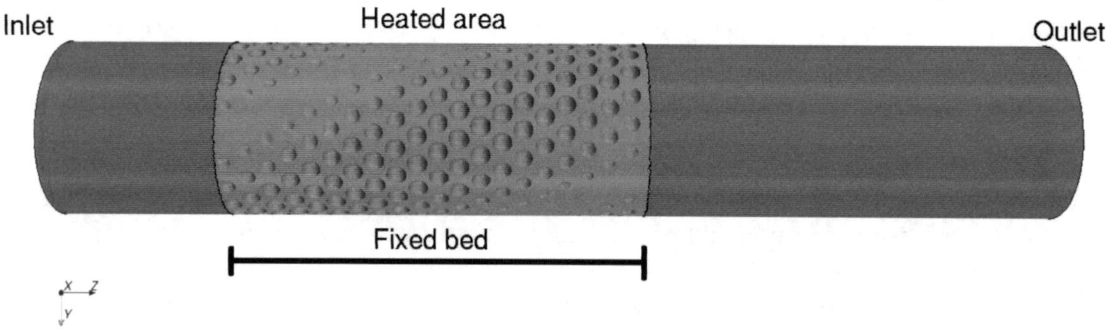

***Figure 2**: Reactor geometry overview: flow direction is from left to right. The red zone marks the heating area (constant temperature T_{wall} = 373 K) as well as the fixed bed. The tube to particle diameter ratio is D/d_p = 8.8.*

on the wall structure of reactor R1, but the fixed bed is newly generated with DEM. Reactor R3 is based on the wall structure of R1 but all spheres whose centroids were within the cutting cylinder are removed.

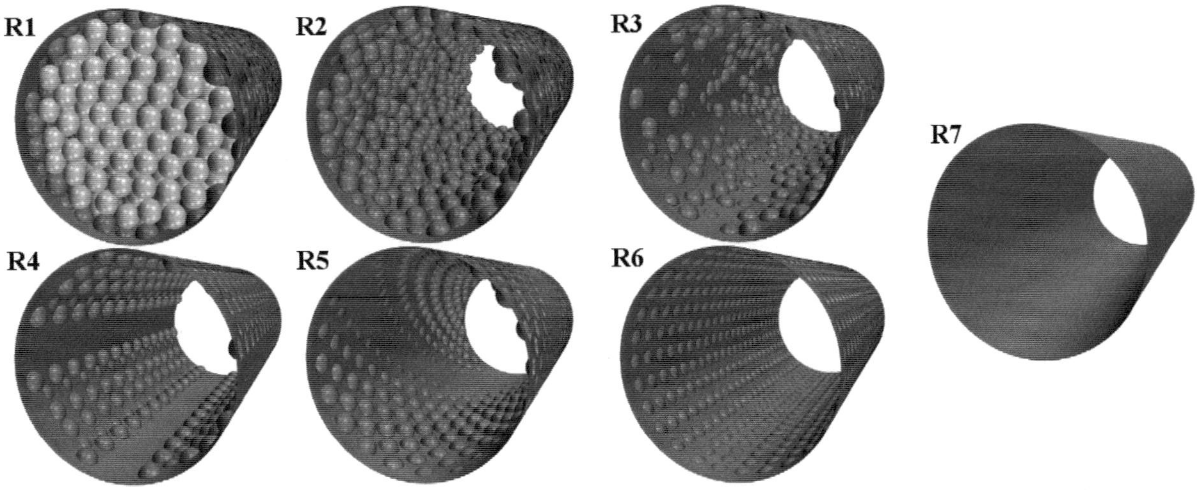

Figure 3: Investigated wall structure overview: R1 – ideal; R2 like R1, but refilled; R3 like R2, but spheres belonging to the wall structure are removed if their centroid resides within the reactor tube; R4-R6 different regular arrangements; R7 plain wall.

For the interpretation of the CFD data several surfaces were generated for post processing: For the evaluation of the radial void fraction distribution cylindrical surfaces were used. Each of these surfaces have an offset of $0.02*d_p$ in radial direction starting at the reactor wall. For each of these surfaces the void fraction as well as the area-averaged radial velocity is determined.

For the evaluation of the development of the radial velocity in axial direction planes perpendicular to the main flow direction are introduced, again with an offset of $0.02*d_p$.

The calculation of the heat transfer coefficient U is based on the logarithmic temperature difference ΔT_{ln}

$$\Delta T_{\text{ln}} = \frac{T_{\text{aus}} - T_{\text{ein}}}{\ln\left(\frac{T_{\text{w}} - T_{\text{ein}}}{T_{\text{w}} - T_{\text{aus}}}\right)} \tag{2}$$

and can be calculated according the following equation:

$$U = \frac{\rho_{\text{f}} \cdot u_0 \cdot c_{\text{p,f}} \cdot \frac{\pi}{4} \cdot D^2 \cdot (T_{\text{aus}} - T_{\text{ein}})}{\pi \cdot D \cdot L \cdot \Delta T_{\text{ln}}} \tag{3}$$

with the fluid density ρ_f, the superficial verlocity u_0, the heat capacity of the fluid $c_{p,f}$, the reactor diameter D and length L and the outlet, inlet and wall temperature T_{outlet}, T_{inlet} and T_{wall}, respectively.

Results

Radial void fraction distribution

In Figure 4 the radial void fraction distributions for the investigated wall structures R1 – R7 are depicted. Reactor R1 with a homogeneous void fraction distribution of ε =0.4 and reactor R7 with the plane wall and therefore the most oscillating void fraction distribution serve as benchmarks. The reactor R7 shows the well-known oscillating and damped radial void fraction distribution starting with a value of $\varepsilon(z^*) = 1$ at the wall, a first minimum at approximately half a sphere diameter and a first maximum at one sphere diameter. Also the constant void fraction for reactor R1 is as expected because the particles and the wall structure are taken from a larger reactor where the void fraction already reached a constant value. It is conspicuous that for reactor R2 with the same wall structure like R1 but with a newly generated particle bed, the radial void fraction profile is also much more homogeneous compared to reactors R3-R7 but slightly higher than for reactor R1. Reactor R3, R4, R5 and R6 show quite similar radial void fraction profiles, which start at the wall with a lower value due to the wall structures and the minima and maxima of the oscillation are less pronounced. The alignment of the sphere caps, whether regular like for reactor R4 and R5 or random like for reactor R3, does not influence the void fraction profile significantly.

As a first result can be noted that with an appropriate wall structure the radial void fraction distribution can be homogenized significantly.

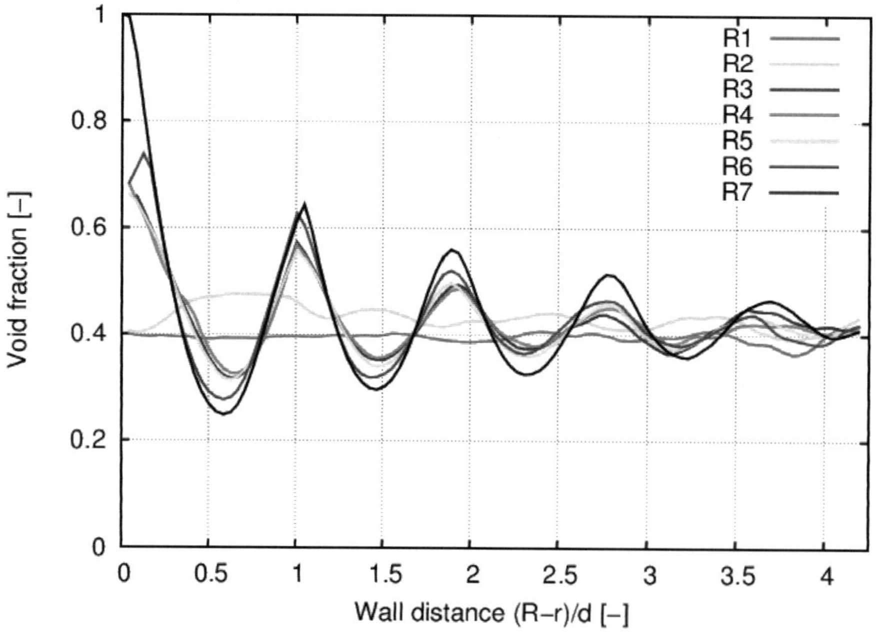

Figure 4: *Radial void fraction profiles for different wall structures.*

Velocity distribution

The quantity which characterizes the extent of lateral mixing in the reactor is the magnitude of the lateral velocity component of the fluid. In Figure 5-a the circumferential and over the bed height averaged magnitude of the radial velocity component u_r normalized with the interstitial velocity $u_\varepsilon = u_0/\varepsilon$ versus the normalized wall distance is depicted. Figure 5-b shows the development of the magnitude of the radial velocity averaged over cross sections normal to the main flow direction in the reactor. For the sake of a clearer presentation of the results the values for reactors R4-R6 are omitted. Analogous to the radial void fraction distribution the velocity distributions of reactors R4-R6 are very similar to that of reactor R3.

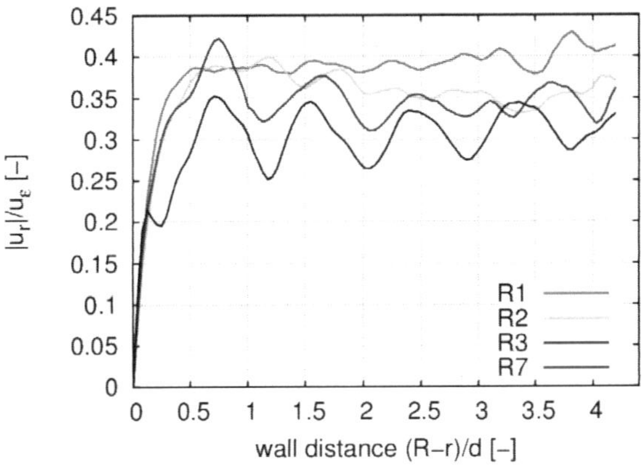

Figure 5a: Averaged (circumferential and over the bed height) magnitude of the radial velocity component ur. A significant increase for reactor with wall structure can be observed compared to reactor R7 (plain wall).

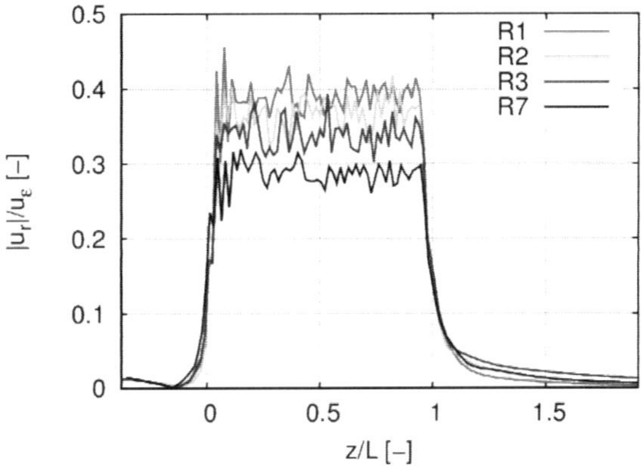

Figure 5b: Axial development of the radial velocity magnitude. Fixed bed starts at z/L = 0 and ends at z/L = 1.

For reactor R7 (plain wall) the radial velocity magnitude oscillates around $u_r/u_\varepsilon = 0.3$ over the wall distance according to the radial void fraction profile. For reactors R1 und R2 such oscillations can no longer be observed: the radial velocity increases from $u_r = 0$ at the wall to its maximum value of approximately $u_r/u_\varepsilon = 0.38$ at half a sphere diameter and stays on this level. The radial velocity magnitude fields of the reactors with regular wall structures R3-R6 can be considered as intermediates between the extreme cases R1/R2 and R7, respectively, both in terms of average value and in terms of extent of oscillations.

The axial development of the radial velocity magnitude, depicted in Figure 5b, starts slightly before the bed (2-3 sphere diameters) caused by the deflection of the flow due to the fixed bed. Within 2-3 sphere diameters it reaches its maximum value and oscillates around a mean value. After leaving the fixed bed at z/L=1 the radial velocity component declines fast to zero. Again the lowest value for the radial velocity magnitude is found for the plain wall reactor R7, the highest values for the reactors R1 und R2 und reactor R3 is somewhere in between.

Based on these results it can be concluded, that with a more homogeneous radial void fraction distribution the flow in the reactor in axial as well as in radial direction is more homogenous. And due to the wall structure the mass flow in the near-wall region is reduced and consequently the mass flow is increased in the inner zone which results in higher radial velocities.

Heat Transport

Based on the conclusions of the last section we now demonstrate the benefit of wall structures on the radial heat transport and the heat transfer from the reactor wall.

Figure 6 shows a comparison of the temperature profile for the reactors R1, R3 and R7. The inlet temperature and velocity are set to T_{in} = 298 K and v_{in} = 0.1 m/s, respectively. The wall temperature T_{wall} = 373 K is assumed to be constant in the heating zone. In the simulations on the left-hand side, series 1, the heat conduction in the particles is neglected while on the right-hand side (series 2) the particles are meshed and modeled with the physical values of glass (λ = 1.4 W/(mK)).

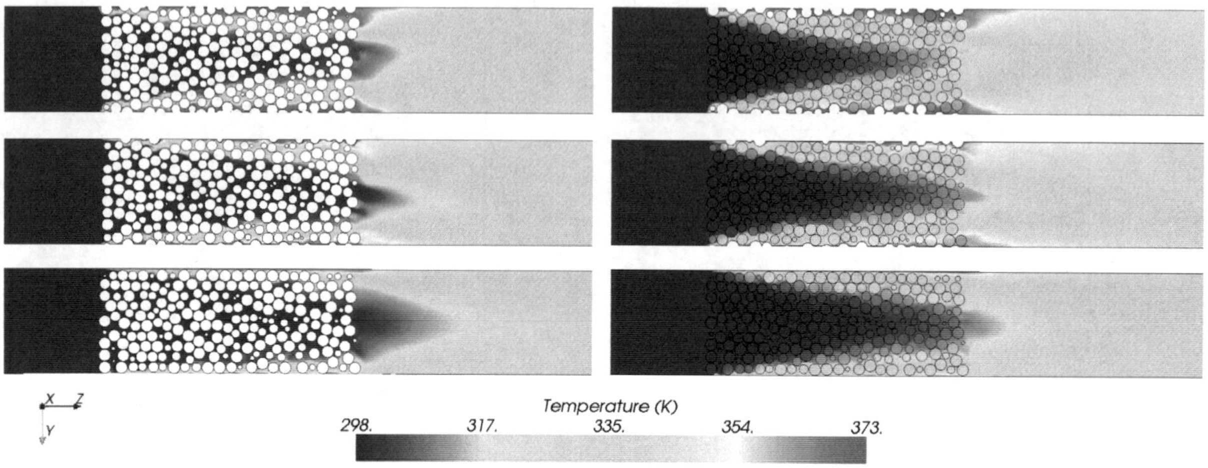

***Figure 6**: Temperature profile for reactor R1 (top), R3 (middle) and R7 (bottom) at Re_p = 158 and a constant heating temperature of T_{wall} = 373 K without (left side) and with (right side) conducting particles.*

Comparing the simulations of series 1, it can be clearly seen that the temperature field is influenced significantly by the wall structure. For the plain wall reactor R7 mainly the temperature in the near-wall region is increased and the bed center is barely influenced. For reactor R3 larger zones are affected while the highest effects can be determined for reactor R1. This can be attributed directly to increased radial velocity due to the wall structure. Similar results can be found for the simulation series 2, although the solid heat conduction now leads to an additional heat flux towards the reactor axis and larger zones show an elevated temperature.

To make the results quantitatively comparable, the global heat transfer coefficient U has been calculated based on the equation (3). Because of the enlarged heat transfer surface caused by the additional surface of the wall structure, the heat transfer coefficient multiplied by the heat transfer surface is also given (U*A). This variable is – from the engineering point of view - more meaningful because it shows directly how much heat can be transferred per unit reactor length.

The results displayed in Figure 7 show that in cases where the convective transport dominates the heat transport (marked as 'without solid') an increase of U*A of 26% to 51% compared to the plain wall reactor R7 is found for reactor R1 and R2, respectively. For reactor R3 the increase is very small.

For the simulations which consider heat transport in the particles all reactors with wall structure show an increase in U*A compared to reactor R7. For reactor R1 and R2 the increase is about 33% and 45% respectively while for R3 it is only 13%. The increase compared to the simulations neglecting heat transport in the solid phase is caused by the additional radial heat conduction in the particles.

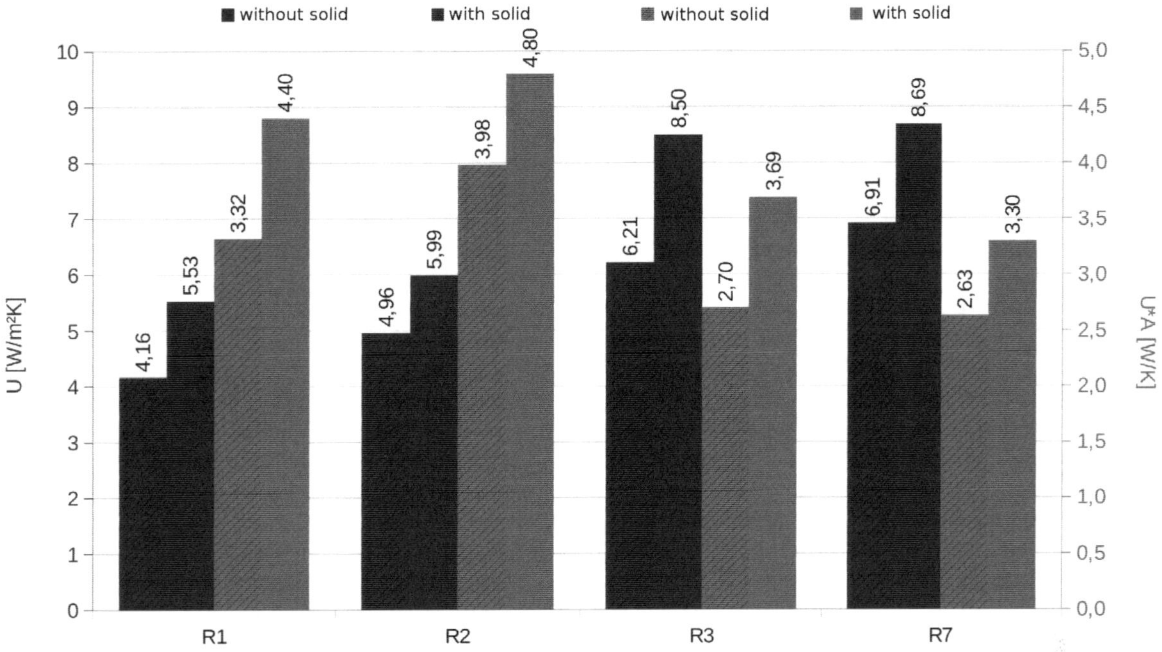

***Figure 7**: Comparison of the heat transfer coefficient U and the product of heat transfer coefficient U and heating area A.*

Summary, Conclusion and Outlook

The general influence of wall structures on the void fraction distribution of a monodisperse spheres packing is investigated mainly numerically, but also experimentally. It can be shown, that with wall structures the radial void fraction distribution tends to be more homogenous and the radial velocity is increased significantly up to 33% compared to its value in a tube without wall structure. The influence of wall structures on the heat transfer is shown. An increase of up to 50% for U*A can be achieved.

Further numerical and experimental investigations are needed to judge the influence of different settings like physical characteristics, higher and lower flow rates and for further optimization of the wall structure arrangement. In addition the impact of wall structures on the performance of catalytic reactions has to be investigated. This is planned for the near future with a detailed reaction mechanism (partial oxidation of methane) which is solved with the specialized stiff ODE solver DARS, which can be directly coupled with the used software STAR-CCM+.

Acknowledgments

The work of Th.Eppinger is part of the Cluster of Excellence "Unifying Concepts in Catalysis" coordinated by the Berlin Institute of Technology. Financial support by the Deutsche Forschungsgemeinschaft (DFG) within the frame work of the German Initiative for Excellence is gratefully acknowledged (EXC314).

References

1. Bai, H., Theuerkauf,J., Gillis,P.A., Witt,P.M. (2009): A coupled DEM and CFD simulation of flow field and pressure drop in fixed bed reactor with randomly packed catalyst particles. Ind. Eng.
2. Cundall, P., Strack, O. (1979): A discrete numerical model for granular assemblies. Geotechnique 29, 47–65.
3. Eppinger, T., Seidler, K., Kraume, M. (2011): DEM-CFD simulations of fixed bed reactors with small tube to particle diameter ratios. Chem. Eng. J. 166(1), 324–331.
4. Guardo, A., Coussirat, M., Angels Larrayoz, M., Recasens, F., Egusquiza, E. (2004): CFD flow and heat transfer in nonregular packings for fixed bed equipment design, Ind. Eng. Chem. Res. 43, 7049–7056.
5. Nijemeisland, M., Dixon, A. G. (2001): Comparison of CFD simulations to experiment for convective heat transfer in a gas–solid fixed bed, Chem. Eng. J. 82, 231-246
6. Ookawara, S., Kuroki, M., Street, D., Ogawa, K. (2007): High-fidelityDEM-CFD modeling of packed bed reactors for process intensification.In: Proceedings of European Congress of Chemical Engineering(ECCE-6), Copenhagen.
7. Winterberg, M.,Tsotsas,E.,Krischke,A.,Vortmeyer,D., (2000): A simple and coherent set of coefficients for modelling of heat and mass transport with and without chemical reaction in tubes filled with spheres. Chem. Eng. Sci. 55 (5), 967–979.
8. Zobel, N., Eppinger, T., Behrendt, F., Kraume, M. (2012): Influence of the wall structure on the void fraction distribution in packed beds, Chem. Eng. Sci. 71, 212-219

(625c) Experimental and CFD Simulation of Methane Steam Reforming Reaction/Deactivation in a Catalytic Fixed-Bed Reactor

Mohsen Behnam[1], Anthony G. Dixon[1], Paul M. Wright[2], Michiel Nijemeisland[2] and E. Hugh Stitt[2], (1)Chemical Engineering Department, Worcester Polytechnic Institute, WORCESTER, MA, (2)Johnson Matthey Catalysts, Billingham, United Kingdom

Modeling of fluid flow, heat transfer and reaction in fixed-beds is an essential part of their design. This is particularly important for highly endothermic reactions in low tube-to-particle diameter ratio (N) tubes. The highly endothermic commercial methane steam reforming (MSR) reaction was studied experimentally and computationally in a low-N fixed-bed reactor at different operating conditions. Experimental results showed strong effect of MSR reaction on temperature contributions inside the catalyst particles. A Computational fluid dynamics (CFD) model of the fixed-bed reactor was developed under the reaction conditions similar to the experiments. The CFD simulation showed very good agreement with experimental data for temperature inside the catalyst particles and in the fluid zone under the MSR reacting conditions.

In addition, a 3D resolved CFD model simulation was extended to study the local details of carbon deposition in which the MSR reactions and deactivation occurred inside the catalytic solid particles. Carbon was formed at low steam to carbon flow rate ratios and high temperature conditions inside the catalyst particles. Carbon was initially deposited fastest at the high-temperature locations close to the surface in the catalyst particles. The results suggested that the shape of catalyst particles can change the potential of carbon formation inside the particles.

(625d) A Plea for Using Lattice Boltzmann Techniques in Simulating Chemical Reactors

Harry E.A. Van den Akker and Mohammad R. Kamali, Dept. of Multi-Scale Physics, Delft University of Technology, Delft, Netherlands

Rather than with the common Finite Volume techniques, the complex interplay of flow and transport phenomena and chemical reactions in single-phase and multi-phase chemical reactors may also be described by means of Lattice Boltzmann (LB) techniques.

In the LB approach, each component or phase is represented by a set of fictitious particles rather than by the usual continuum variables. These fictitious particles, which are allowed to move on a regular lattice with their own discrete velocity distribution functions, carry the macroscopic properties such as density. These particles move and collide with each other as hard spheres resulting in a redistribution in the velocity space such that total mass, momentum and energy in the system are conserved . In the so-called BGK approximation (on the lattice type D3Q18), this redistribution is conceived as a relaxation to an equilibrium Maxwell distribution with a single relaxation. This relaxation time is an important parameter whose role of responding to spatial or temporal variations is equivalent to that of viscosity in the flow of fluids or of diffusion coefficients in species transport

The method we are developing exploits the pseudo-potential multi-component multi-phase LB model due to Shan and Chen in which the interaction between fictitious particles representing the different components is described in terms of the interaction potential functions ψ of the multiple components and a so-called coupling strength Gij. This coupling strength affects the degree of miscibility and solubility of two components, the sharpness of a phase interface as well as the surface tension (in the case of a gas-liquid system), and the wettability of a solid surface (in the case of a gas/liquid/solid contact: see e.g., (Kamali, Gillissen et al. 2011)). The Shan-Chen method can be modified to include more realistic equations of state allowing to deal with phase density ratios of practical interest.

In addition, the Shan-Chen method is capable of dealing with multi-component convection-diffusion problems. Realistic mass transport across a phase interface (with the Schmidt number much bigger than unity on the liquid side of the interface) is obtained by allowing the relaxation time to depend on the local density. In order to incorporate a surface reaction (at a catalytic surface), a modified bounce-back rule – the usual LB technique for mimicking the no-slip boundary condition in flow– was applied.

In general, using LB techniques requires a translation from the conventional continuum variables (velocities, concentration, viscosity, diffusivity, ….) into typical LB variables. More specifically, running successful LB simulations depends on selecting proper values for three types of typical LB variables: the potential functions ψ, the coupling strengths Gij, and the relaxation time. This selection process is especially delicate when two-phase flow and multi-component mass transport have to be simulated.

The idea of exploiting LB techniques for CRE applications is illustrated by applying the technique to the case of an isothermal 1-D Fischer-Tropsch reaction involving a multi-component chemical reaction with phase change at a catalytic surface.

(625e) Fouling of Convection Section Tubes Due to Heavy Feeds in a Steam Cracker

Amit V. Mahulkar, Geraldine J. Heynderickx and Guy B. Marin, Laboratory for Chemical Technology, Ghent University, Ghent, Belgium

Enhanced fouling rates of convection section tubes of a steam cracker is being increasingly observed owing to the tendency of the industrial crackers to crack heavier feeds. CFD simulations are performed to predict the extent of fouling i.e. coke formation in the over-heater tubes of the convection section of a steam cracker. Several orientations of tube geometries of an actual steam cracker were simulated each having 3 passes and each pass being 11 m long of 0.0775 m diameter [1]. Operating conditions that were simulated include inlet droplet diameter in the range 1 to 100 mm and outer wall temperature above and below the boiling point of the highest boiling component in the feed.

The spray flow was simulated using the discrete phase model (Lagrangian approach). Eleven representative chemical species were selected to represent the entire feed composition in terms of its physico-chemical properties. Evaporation of species from the multicomponent droplets while flowing through the tube was also included in the simulations. The droplet interactions with the wall (splash, limited splash, rebound and stick) were simulated using the splashing regime map of Bai et.al. (2002)[2],adapted to the present conditions where needed. Movement of the liquid deposited on the wall was simulated using the Eulerian wall film model. The latter also took into of account evaporation of liquid, and generation of droplets due to shearing of the liquid wall film of already deposited droplets. The dry and wet patches on the tube wall, owing to boiling in the liquid film, is also accounted for, along with changes in the heat transfer characteristics of the wall. The quantity and the composition of the liquid on the wall film is determined using a coking rate model based on the coking model of Wiehe (1993)[3].

It was seen that the coking rates are dominantly controlled by the outer wall temperature in the convection section. Higher coking rates are observed for outer wall temperatures lower than the boiling point of the feed. For outer wall temperature higher than the boiling point of the deposited liquid, coking was seen to occur only in the cold zones (inlet bends) of the tubes. The variation in inlet droplet diameters was seen to have a significant effect on the extent of liquid deposition on the wall. The effect of variation in inlet droplet diameters, the shell side outer wall temperature and the tube geometry on the amount of liquid deposited on tube wall and coking rates are discussed. The trend in coke formation rates and its effect on a change in heat transfer and pressure drop characteristics of the tubes is studied for over an operating period of 1 year.

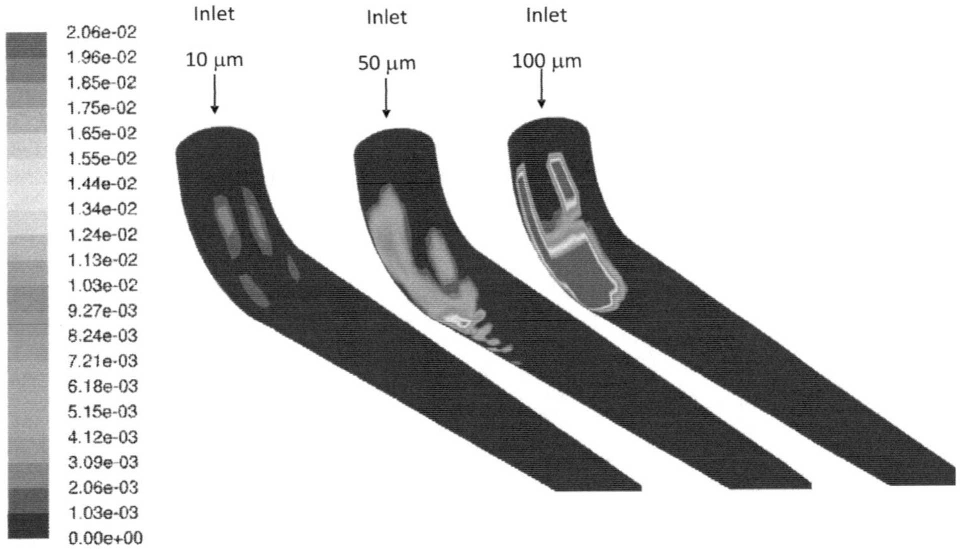

Fraction of incoming liquid deposited at the inlet bend for different inlet droplet diameters

References

1. S. De Schepper, G. J. Heynderickx, G. B. Marin, (2010), Modeling the coke formation in the convection section tubes of a steam cracker, Ind. Eng. Chem. Res., 49, 5752-5764.

2. C.X. Bai, H. Rusche & A.D. Gosman, (2002), Modeling of gasoline spray impingement, Atomization and Sprays, 12, 1-27

3. I. Wiehe, A phase separation kinetic model for coke formation. Ind. Eng. Chem. Res. 32 (1993) 2447-2454

Development of realistic liquid equations of state for multi-phase CFD modelling

David J. Robbins, University of Cambridge, United Kingdom (djr57@cam.ac.uk)
Lynn F. Gladden, University of Cambridge, United Kingdom (lfg1@cam.ac.uk)
R. Stewart Cant, University of Cambridge, United Kingdom (rsc10@cam.ac.uk)

Abstract

Several equations of state (EOS) have been incorporated into a novel algorithm to solve a system of multi-phase equations in which all phases are assumed to be compressible to varying degrees. The EOSs are used to both supply functional relationships to couple the conservative variables to the primitive variables and to calculate accurately thermodynamic quantities of interest, such as the speed of sound. Each EOS has a defined balance of accuracy, robustness and computational speed; selection of an appropriate EOS is generally problem-dependent. This work employs an AUSM$^+$-up method for accurate discretisation of the convective flux terms with modified low-Mach number dissipation for added robustness of the solver.

In this paper we show a newly-developed time-marching formulation for temporal discretisation of the governing equations with incorporated time-dependent source terms, as well as considering the system of eigenvalues that render the governing equations hyperbolic.

Introduction

The need to perform accurate simulations of multi-phase flows is of great importance throughout chemical engineering for the prediction of mass, momentum and heat transfer and to provide parameters for design, scale-up and optimisation of process equipment.

Computational fluid dynamics (CFD) is a simulation tool that solves the governing flow equations over a numerically discretised domain in order to predict the state of flow in the system. However, multi-phase systems are complicated to solve in comparison to single-phase systems as they introduce problems that are both physical and mathematical in origin.[1] Modern commercial solvers generally employ one of three models: the mixture model,[2] volume-of-fluid model[3] or the Euler-Euler model.[4] These solvers generally assume incompressibility of the two phases and are also typically pressure-based, imposing a Laplacian condition on the pressure.[5] In this work we employ equations developed from time-averaging of the instantaneous field equations coupled with equations of state for both phases to describe a density-based system. An advantage of this approach is the accurate prediction of thermodynamic quantities and the relative ease of parallelisation,[6] a key factor when considering larger numerical domains.

This paper presents a discussion of the selection of an appropriate EOS for use in compressible two-fluid equations and modifications of an algorithm developed in previous work.[7] The work presented is part of an on-going effort to extend an unstructured single-phase code to a robust multi-phase code in order to predict two-phase flow in trickle-bed reactors. All simulations presented in this paper have been generated using a one-dimensional inviscid code, PULSAR-lite, which was written in FORTRAN to test different the numerical schemes and closure schemes available.

Numerical Method

Governing equations

The time-averaged governing equations in column vector form are given by:

$$\frac{\partial \boldsymbol{W}_i}{\partial t} + \frac{\partial \boldsymbol{F}_i}{\partial x} = \boldsymbol{S}_i^d + \boldsymbol{S}_i^{nv}, \tag{1}$$

where i is the phase index and the column vectors of conservative variables, convective flux variables, differential and non-differential source terms are defined respectively as:

$$\boldsymbol{W}_i = [\alpha\rho, \alpha\rho\boldsymbol{v}, \alpha\rho E]_i^T \tag{2}$$

$$\boldsymbol{F}_i = [\alpha\rho\boldsymbol{v}, \alpha\rho\boldsymbol{v}\boldsymbol{v} + \alpha p, \alpha\rho\boldsymbol{v}H]_i^T \tag{3}$$

$$\boldsymbol{S}_i^d = [0, p\frac{\partial \alpha}{\partial \boldsymbol{x}} + F^{nv}, -p\frac{\partial \alpha}{\partial t} + \boldsymbol{v}^{int}F^{nv}]_i^T \tag{4}$$

$$\boldsymbol{S}_i^{nd} = [0, \alpha\rho\boldsymbol{g} + F^D, \alpha\rho\boldsymbol{v}\boldsymbol{g} + \boldsymbol{v}^{int}F^D]_i^T, \tag{5}$$

where α is the phase volume fraction, ρ the density, $\boldsymbol{v}$ the velocity vector, E the total energy, p the pressure (shared between both phases, such that $p_1 = p_2 = p$), H the total enthalpy, F^{nv} the non-viscous differential source terms (specifically the interfacial pressure), $\mathbf{v}^{\text{int}}$ the interphase velocity, $\boldsymbol{g}$ the gravitational acceleration vector and F^D the interphase drag force.

It is noted that the sharing of pressure between phases specifies a functional relationship to what is an under-specified system, but results in an ill-posed problem.[1] This is alleviated by the use of the interfacial pressure term F^{nv}, relaxing the criterion somewhat for equal pressure between both phases.[8] In previous work,[7] it was shown that failure to include the interfacial pressure term leads to unphysical oscillations in solution profiles, whilst inclusion of this term led to smooth, physical profiles that matched analytical solutions.

Numerical Method

The convective fluxes are discretised using a multi-phase extension of the AUSM$^+$-upwind scheme.[9] Considering a computational cell *m*, the basic premise of the AUSM scheme is to split the convective flux through a face *k* demarcating the cell *m* into convective and pressure terms:

$$\boldsymbol{F}_i = \dot{m}_i\phi_i + \alpha_i\boldsymbol{p}, \tag{6}$$

where $\phi = [1, \boldsymbol{v}, H]_i^T$ is the column vector of convected variables and $\boldsymbol{p} = [0, p, 0]^T$ is the column vector of pressure. The phase mass flux is calculated by consideration of the Mach-number of the flow at the face, $M_f = \boldsymbol{v}_f / c_f$ (where c_f is the speed of sound at the face), determined as a polynomial function of the Mach numbers of the cells on either side of the face. This allows the phase mass flux to be determined by an upwinding method:

$$\dot{m}_i = \alpha_i\rho_i\boldsymbol{v} = \left\{c_f\left[\frac{\alpha_L\rho_L}{2}(M_f + |M_f|) + \frac{\alpha_R\rho_R}{2}(M_f - |M_f|)\right]\right\}_i, \tag{7}$$

where subscripts L and R imply values taken from the left- and right-hand cells of the face under consideration. The value of ϕ_i is selected by upwinding based on the direction of the mass flux. This system is dissipated using low-Mach number terms in order to scale the Mach number polynomials correctly for low-Mach number flows. The full detail of the polynomial schemes and the form of the low-Mach number dissipation are listed elsewhere.[7,9]

One important source term to note is the $\boldsymbol{v}^{int}F^{nv}$ term. The choice of interphase velocity is arbitrary; other studies choose to employ the velocity of the phase under consideration. In this work a blended velocity is chosen (specifically, the arithmetic average); it is shown in Appendix A that the selection of interphase velocity bears no impact on the eigenstructure of the system and therefore does not impact the hyperbolic character of the interfacial pressure-enhanced multi-phase equations.

For temporal discretisation, the source term containing the temporal derivative of the volume fraction poses a challenge. In previous work,[7] this was treated as an explicit source term and was discretised by storing the value at the previous time-step for a forward-Euler method, or pseudo-time step for Runge-Kutta schemes. In this work, it will be incorporated into the conservative variable flux vector and treated implicitly as part of the time-marching. This term includes a pressure term that is frozen at either the current time step (i.e. the value is known during the update algorithm) or at the future time step (determined implicitly during the update algorithm). Two Runge-Kutta (RK) schemes have been incorporated into PULSAR-lite; the first is an explicit time-variable diminishing (TVD) Runge-Kutta scheme,[10] which for an rth-order scheme (ERK-*r*) for cell *m* is:

$$\boldsymbol{W}_m^{(0)} = \boldsymbol{W}_m^{(t)} \tag{8}$$

$$\boldsymbol{W}_m^{(i)} = \sum_{k=0}^{i-1}\left[\beta_{ik}\boldsymbol{W}_m^{(k)} + \gamma_{ik}\Delta t\mathcal{L}\left(\boldsymbol{W}_m^{(k)}, t + \lambda_k\Delta t\right)\right] \tag{9}$$

$$\boldsymbol{W}_m^{(r)} = \boldsymbol{W}_m^{(t+\Delta t)}, \tag{10}$$

where β_{ik} , γ_{ik} and λ_k are order-dependent constants available in the original paper, Δt is the global minimum time-step determined by an acoustic-signals algorithm[11] and $\mathcal{L}(\boldsymbol{W}, t)$ is the spatial operator acting on the vector $\boldsymbol{W}$ at time t. A second method incorporated is a multi-stage Runge-Kutta scheme; the four-stage scheme (MS-RK4) is implemented for the conservative flux vector as:[12]

$$\boldsymbol{W}_m^{(0)} = \boldsymbol{W}_m^{(t)}$$

$$\boldsymbol{W}_m^{(i)} = \boldsymbol{W}_m^{(0)} - \theta_i\Delta t\mathcal{L}\left(\boldsymbol{W}_m^{(0)}, t + \theta_i\Delta t\right)$$

$$\boldsymbol{W}_m^{(4)} = \boldsymbol{W}_m^{(t+\Delta t)},$$

for which the stage coefficients are defined simply as $\theta_i = \{0, 0.25, 0.50, 1.00\}$. The multi-stage RK scheme is of lower-order accuracy than the ERK scheme, but is generally more robust in particularly challenging flows where large gradients of the primitive variables drive large changes in the conservative fluxes with respect to time.

Equations of State

The equations of state employed in PULSAR-lite provide functional relationships for density $\rho = \rho(p, T)$ and internal energy $e = e(p, T)$. The speed of sound must also be given as a function of pressure and temperature. The EOSs incorporated can be subdivided into four principal groups:

- *Empirical EOSs*: IAPWS-IF97 industrial formulation for water, simple polynomial correlations;
- *Developmental EOSs*: stiffened-gas, Tait's equation;
- *Cubic EOSs*: Peng-Robinson (original & modified); Soave-Redlich-Kwong;
- *Principle of corresponding-states*: Yamada-Gunn correlation, Laplace-Newton equation, Ruzicka-Domalski group-contribution equations.

All of these EOSs are described in detail in previous work and standard thermodynamic textbooks, to which the interested reader is referred.[7,13] Of particular interest for this work are the modified Peng-Robinson EOS (PR-EOS) and the corresponding-states principles-based correlations (CSP-EOS). The PR-EOS has been modified by coupling the Ruzicka-Domalski heat capacity polynomials to the original PR-EOS and by adding a correction term for the molar volume based on the isothermal bulk modulus of the liquid under consideration. This yields a more robust EOS that is also far more accurate than the original equation, especially when dealing with strongly polar or non-hydrocarbon liquids such as water. The CSP-EOS is a coupling of three thermodynamic correlations based on critical properties, applicable to all liquids, and yields surprisingly accurate values of density, speed of sound and heat capacity, with errors in the range of 0.2-3.5% at standard temperature and pressure for properties of *n*-hexane and water.

Primitive variable update algorithm

By incorporating the time derivative-containing source term into the conservative variables, the conservative energy variable at the future time step becomes $(\alpha\rho E + \alpha p^*)_i^{t+\Delta t}$, where p^* is fixed during the time-marching procedure and can be defined either as pressure at the current time step, p^t, or pressure at the future time step $p^{t+\Delta t}$. This makes no difference to the overall algorithm, but it must be noted that if it is taken to be at the future time-step then a forward-Euler step is required to initialise the fluxes at the beginning of the simulation. To decode the energy conservative variable, the following algorithm determines the position of the state of flow on the liquid (p, T_l) and gas (p, T_g) pressure-temperature surfaces by:

1. Reform velocities of phase i from $(\alpha\rho v)_i^{t+\Delta t}/(\alpha\rho)_i^{t+\Delta t}$;
2. Initialise temperature and pressure guess at the future time step;
3. Determine liquid density and enthalpy using the guessed temperature and pressure, and resolve guessed volume fraction at the future time step with $(\alpha\rho)_i^{t+\Delta t}/\rho_i^{guess}$;
4. Form guessed total liquid energy from the liquid enthalpy determined in (3) and substitute known values to find the root T_l of the equation $F(T_l) = 0 = \left[(\alpha\rho E + \alpha p^*)_i^{t+\Delta t}\right]^{guess} - \left[(\alpha\rho E + \alpha p^*)_i^{t+\Delta t}\right]^{RK}$ - this is solved by a Newton-Raphson procedure;
5. Repeat steps (3) and (4) for the gas phase to determine T_g; a Newton-Raphson solver was found to be divergent for $F(T_g)$ and thus a secant solver was used for the gas;
6. With the located liquid and gas temperatures for the guessed pressure, update the pressure guess by finding the root p of the equation $G(p) = 0 = \left(1 - \frac{(\alpha_g\rho_g)^{t+\Delta t}}{\rho_g^{guess}}\right)\rho_l^{guess} - (\alpha_l\rho_l)^{t+\Delta t}$; this is solved with a Newton-Raphson procedure unless the pressure derivative of density is particularly complex, in which case a secant method is usually faster and has a smaller domain of dependence of the initial guess;
7. With the new pressure guess, go back to step (3). Exit loop once $G(p)$ has reached a desired tolerance; in this study the tolerance is set to 10^{-10}.

Results and Discussion

Only a brief selection of results is presented in this paper; the interested reader is directed to previous work[7,14] for full dissemination of results for the water-air and *n*-hexane-air systems. Ransom's faucet problem[15] has been employed to test the new implicit-time source term algorithm; in this problem a liquid column accelerates under the action of gravity, causing gas entrainment at the outlet and an eventual quasi-steady state solution in which the liquid profiles achieve steady-state. Analytical solutions are available for the volume fractions and liquid velocity if it is assumed that interphase forces are neglected; in this study a transient state at $t = 0.5$ s has been used as it provides a discontinuity in volume fraction, adding to the numerical challenge.

In Figure 1a) the solution profile for liquid mass flux $m_l = \alpha_l \rho_l v_l$ is shown in comparison to an analytical solution for the water-air system. The analytical solution employs the highly-accurate IAPWS-IF97 formulation[16] to determine liquid thermodynamic properties and uses explicit source-term marching rather than incorporating the time-dependent source term into the time-marched variables. The second- and third-order ERK and multistage RK4 scheme employ the modified Peng-Robinson scheme. It can be seen that a sharp resolution of the liquid mass flux discontinuity is given (for the domain of 2000 cells) employing the present algorithm with the upwinded flux-splitting scheme. This sharpness will be critical in applications in which resolution of the fluid interface is important, such as the flows present in packed-bed reactors or falling-film columns. No discernible difference is noted between the solution profiles of the different time discretisation schemes; it was shown in previous work[7] that a simple forward-Euler scheme was inadequate for resolving stratified-type flow, whilst the ERK schemes generated a smooth, physical profile.

Figure 1b) shows the gas temperature drop over the domain, $\delta T_g = T_g - T_{(g,0)}$ for the new implicit scheme under three different temporal schemes compared with the IAPWS-reference system employing explicit source term discretisation. It can be seen immediately that the previous method resulted in a small instantaneous drop in gas temperature at the inlet, for which a gas temperature of 298 K is prescribed. This is physically unrealistic. In the present method, this error is reduced from 0.35 K to 0.075 K and the solution profile is smoother and oscillation-free. It is noted that the drop of 0.2 K in the centre of the profile corresponds to a discontinuity in volume fraction, where the transient mixed flow becomes predominantly liquid. Other solution variables such as velocity and pressure yield the same profiles as for the explicit time-dependent source term formulation and are thus not shown.

Figure 1. Solution profiles for Ransom's faucet test case with a water-air system. a) Liquid mass flux (m_l) profile for faucet problem compared with analytical solution, b) temperature drop through domain relative to inlet temperature (T_l) compared with explicit source term reference solution.

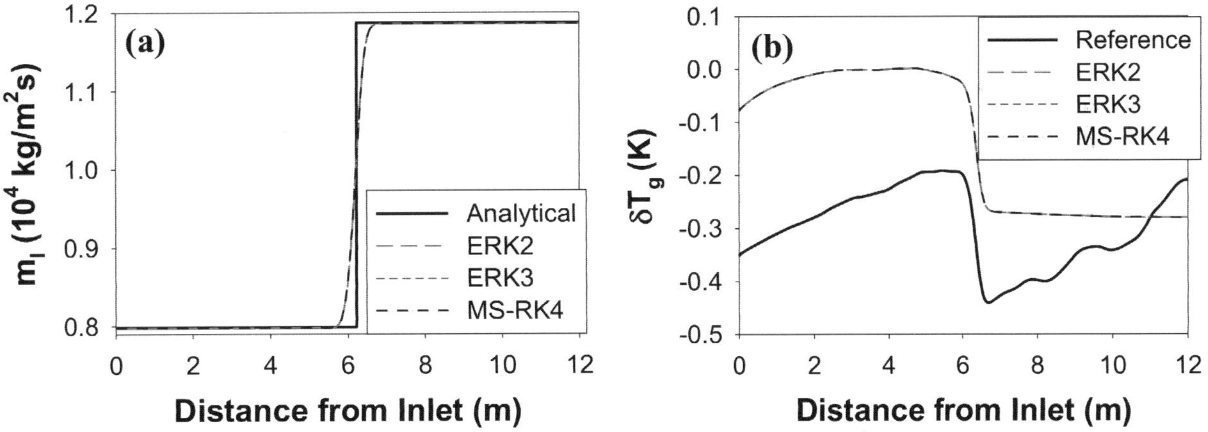

In Table 1 predictions of the water density and speed of sound are provided for the correlations based on the principle of corresponding-states and the cubic EOSs. It can be seen that the correlations based on the principle of corresponding-states (employing the Newton-Laplace equation for speed of sound) are remarkably accurate for a generalised set of correlations, with an error of less than 1% for both properties. Cubic EOSs are notoriously poor for estimating properties of strongly polar liquids such as water and this is reflected in the errors of the original PR-EOS and enthalpy-coupled SRK. The modified PR-EOS, employing a volume fraction correction and assumed ratio of specific heats for the speed of sound prediction, performs much better. Accurate determination of the speed of sound is important in order for the Mach number splitting polynomials in the AUSM$^+$-up scheme to be accurate and that the correct amount of low Mach number dissipation is added to the system. The accuracy of the predicted values for liquid density and speed of sound has been evaluated over pressure ranging from 1-200 bar under temperatures of 298-308 K in previous work.[14] For the water system it was shown that the maximum error in the CSP-EOS predictions was 1.7% and 3.9% for the density and speed of sound respectively; these errors were 0.5% and 2.8% for the modified PR-EOS. For *n*-hexane the maximum errors in the CSP-EOS predictions are 2.6% and 11.7% for density and speed of sound respectively, and 1.6% and 3.2% for the modified PR-EOS. The CSP-EOS error for speed of sound originates from the use of an isothermal bulk modulus in the Newton-Laplace equation; at higher pressures and elevated temperatures the isothermal assumption generates a larger source of error.

	Density, ρ (kg/m³)	*Speed of Sound, c (m/s)*
Reference	997.08	1497.75
Corresponding-states	997.88	1487.05
Peng-Robinson (original)	848.70	3142.88
Peng-Robinson (modified)	997.09	1496.72
Soave-Redlich-Kwong (modified)	755.74	2963.79

Table 1. Comparison of density and speed of sound predicted by the cubic EOSs and corresponding-states principles for water to the reference IAPWS-IF97 formulation at 1 bar, 298 K.[16]

Conclusions

An algorithm has been developed that incorporates the time-dependent source term of the averaged multi-phase field equations into the time-marching procedure. This algorithm employs realistic liquid equations of state to eliminate the assumption of incompressibility of the phases and a density-based solver to couple the pressure and velocity. It has also been shown mathematically that the velocity selected for the source term of energy generated by motion of the interface does not impact the hyperbolicity of the system and is therefore a generic selection; incorporating motion from both phases is more physically grounded due to the motion of the fluid-fluid interface being imparted by the motion of both fluids.

It was noted that the present algorithm resulted in a smaller error in the gas temperature field than an algorithm in which the time-derivative source term was treated explicitly, whilst simulations of other fluid profiles (for which analytical solutions are available) remain consistent. Minimising temperature error is important as this influences other temperature-dependent variables such as gas density and speed of sound. Future work will encompass addition of further source terms such as phase viscosity and surface tension, and the algorithm will be implemented into a full three-dimensional unstructured solver in order to simulate realistic systems.

Acknowledgements

This work has been funded by the CASTech project (EPSRC reference EP/G011397/1) and the financial support of the EPSRC and Johnson Matthey is greatly appreciated.

References

1. Stewart HB, Wendroff B. Two-Phase Flow: Models and Methods. *J. Comp. Phys.* 1984;56:363-409.
2. Soo S. On One-Dimensional Motion of a Single Component in Two-Phases. *Int. J. Multi. Flow.* 1976;3:79-82.

3. Hirt C and Nichols B. Volume of Fluid (VOF) Method for the Dynamics of Free Boundaries. *J. Comp. Phys.* 1975;39:201-225.
4. Ishii M. Thermo-Fluid Dynamics of Two-Phase Flow (2nd edition). New York: Springer, 2010.
5. Zhang X, Ni S, He G. A Pressure-Correction Method and its Applications on an Unstructured Chimera Grid. *Comput. Fluids.* 2008;37:993-1010.
6. Kwak D, Kiris C. CFD for incompressible flows at NASA Ames. *Comput. Fluids.* 2009;38:504-510.
7. Robbins DJ, Cant RS, Gladden LF. Development of accurate, robust liquid equations of state for multi-phase CFD simulations with a modified AUSM$^+$-up scheme. *Comput. Fluids.* (submitted for review)
8. Zanotti AL, Méndez CG, Nigro NM, Storti M. A Preconditioning Mass Matrix to Avoid the Ill-Posed Two-Fluid Model. *J. Appl. Mech-T. ASME.* 2007;74:732-740.
9. Liou MS. A Sequel to AUSM, Part II: AUSM$^+$-up for All Speeds. *J. Comp. Phys.* 2005;214:137-170.
10. Shu C, Osher S. Efficient Implementation of Essentially Non-Oscillatory Shock-Capturing Schemes, II. *J. Comp. Phys.* 1989;83:32-78.
11. Paillère H, Corre C, García Cascales J. On the extension of the AUSM$^+$ scheme to compressible two-fluid models. *Comput. Fluids.* 2003;32:891-916.
12. Blazek J. Computational Fluid Dynamics: Principles and Applications (1st edition). Oxford: Elsevier Science Ltd, 2001.
13. Smith J, Van Ness H and Abbott M. Introduction to Chemical Engineering Thermodynamics (7th edition). New York: McGraw-Hill, 2005.
14. Robbins DJ, Cant RS, Gladden LF. Novel liquid equations of state for CFD simulations of compressible multi-phase flow. *Int. J. Multi. Flow.* (submitted for review)
15. Ransom V. Numerical benchmark tests, in: Hewitt G, Delhaye J, Zuber N (eds.), Multiphase Science and Technology (vol. 3). Hemisphere Publishing Corporation, 1987.
16. Wagner W, Cooper J, Dittmann A, Kijima K, Kretzschmar H, Kruse A, Mares R, Oguchi K, Sato H, Stocker I, Sifner O, Takaishi Y, Tanishite I, Trubenbach J, Willkommen T. The IAPWS industrial formulation 1997 for the thermodynamic properties of water and steam. *Trans. ASME.* 2000;122:150-182.

Appendix A: Eigensytem

We note that the system of equations in (1) can be re-written in one dimension as:

$$\boldsymbol{A}\frac{\partial \boldsymbol{P}}{\partial t} + \boldsymbol{B}\frac{\partial \boldsymbol{P}}{\partial x} = 0, \tag{11}$$

where $\boldsymbol{P} = \left[p, \alpha_g, v_l, v_g, S_l, S_g\right]^T$ is the column vector of primitive variables and the entropy S_i has been used as the primitive variable for energy. The matrix $\boldsymbol{A}$ incorporates both the conservative variable Jacobian matrix and the time-derivative source terms; the matrix $\boldsymbol{B}$ likewise incorporates the convective flux Jacobian and the spatial derivative source terms. After considerable simplification, and assuming that $v^{int} = \phi v_g + (1-\phi)v_l$, these matrices may be written as:

$$\boldsymbol{A} = \begin{pmatrix} \alpha_l/c_l^2 & -\rho_l & 0 & 0 & 0 & 0 \\ \alpha_g/c_g^2 & \rho_g & 0 & 0 & 0 & 0 \\ \alpha_l v_l/c_l^2 & -\rho_l v_l & \alpha_l\rho_l & 0 & 0 & 0 \\ \alpha_g v_g/c_g^2 & \rho_g v_g & 0 & \alpha_g\rho_g & 0 & 0 \\ \alpha_l S_l/c_l^2 & -\rho_l S_l & 0 & 0 & \alpha_l\rho_l & 0 \\ \alpha_g S_g/c_g^2 & \rho_g S_g & 0 & 0 & 0 & \alpha_g\rho_g \end{pmatrix} \tag{12}$$

$$
\boldsymbol{B} = \begin{pmatrix}
\alpha_l v_l / c_l^2 & -v_l \rho_l & \alpha_l \rho_l & 0 & 0 & 0 \\
\alpha_g v_g / c_g^2 & v_g \rho_g & 0 & \alpha_g \rho_g & 0 & 0 \\
\alpha_l + \alpha_l u_l^2 / c_l^2 & p - \delta_p - \rho_l v_l^2 & 2\,\alpha_l \rho_l v_l & 0 & 0 & 0 \\
\alpha_g + \alpha_g u_g^2 / c_g^2 & \delta_p - p + \rho_g v_g^2 & 0 & 2\,\alpha_g \rho_g v_g & 0 & 0 \\
\alpha_l v_l S_l / c_l^2 & (\phi v_g + (1-\phi) v_l)\delta_p - \rho_l v_l S_l & \alpha_l \rho_l S_l & 0 & \alpha_l \rho_l v_l & 0 \\
\alpha_g v_g S_g / c_g^2 & -(\phi v_g + (1-\phi) v_l)\delta_p - \rho_g v_g S_g & 0 & \alpha_g \rho_g S_g & 0 & \alpha_g \rho_g v_g
\end{pmatrix} \tag{13}
$$

where δ_p is the interfacial pressure term. For a hyperbolic system, it is required that the eigenvalues of the matrix $\boldsymbol{A}^{-1}\boldsymbol{B}$ are all real. This matrix is:

$$
\boldsymbol{A}^{-1}\boldsymbol{B} = \xi \begin{pmatrix}
\alpha_l \rho_g v_l c_g^2 + \alpha_g \rho_l v_l c_g^2 & \rho_l \rho_g c_l^2 c_g^2 (v_g - v_l) & \alpha_l \rho_l \rho_g c_l^2 c_g^2 & \alpha_g \rho_l \rho_g c_l^2 c_g^2 & 0 & 0 \\
(v_g - v_l)\alpha_l \alpha_g & \alpha_l \rho_g v_g c_g^2 + \alpha_g \rho_l v_l c_l^2 & -\alpha_l \alpha_g \rho_l c_l^2 & \alpha_l \alpha_g \rho_g c_g^2 & 0 & 0 \\
1/\xi \rho_l & \Delta p / \alpha_l \rho_l \xi & v_l/\xi & 0 & 0 & 0 \\
1/\xi \rho_g & \Delta p / \alpha_g \rho_g \xi & 0 & v_g/\xi & 0 & 0 \\
0 & (\phi v_g + (1-\phi) v_l)\delta_p / \alpha_l \rho_l \xi & 0 & 0 & v_l/\xi & 0 \\
0 & -(\phi v_g + (1-\phi) v_l)\delta_p / \alpha_g \rho_g \xi & 0 & 0 & 0 & v_g/\xi
\end{pmatrix} \tag{14}
$$

with $\xi = \left(\alpha_l \rho_g c_g^2 + \alpha_g \rho_l c_l^2\right)^{-1}$ and $\Delta p = (p - \delta_p)$. We note that the fifth and sixth diagonal elements give the first two eigenvalues v_l and v_g immediately; the zeros present in the fifth and sixth columns also eliminate all presence of the interphase velocity terms in the remaining eigenvalues when considering the determinant of the matrix $\boldsymbol{A}^{-1}\boldsymbol{B}$. The inner 4×4 matrix yields the final four eigenvalues, which may be condensed into a simpler form as:

$$
(\phi + \delta)^2(\phi - \delta)^2 + \chi_a(\phi + \delta)^2 + \chi_b(\phi - \delta)^2 + \chi_c = 0, \tag{15}
$$

with $\chi_a = \alpha_g \Delta p / c_g^2 - \alpha_l \rho_g$, $\chi_b = -(\alpha_l \Delta p / c_l^2 + \alpha_g \rho_l)$, $\chi_c = \Delta p(\alpha_l - \alpha_g)$, $\phi = (\lambda - \frac{u_l + u_g}{2})/r$, $\delta = (u_l - u_g)/2r$ and $r = \left(a_l^2 a_g^2 / \xi\right)^{1/2}$. It is noted that the quartic polynomial in λ is monic, and therefore can be redefined in resolvent cubic form; if the resolvent cubic has three real negative roots then the quartic equation has four real roots. The resolvent cubic itself can be recast in depressed form; the final condition yields that the discriminant of the depressed cubic must be non-positive. The final inequality for hyperbolicity is too large to be included here; by substituting appropriate numbers (e.g. $\rho_l = O(10^3)$) one may conduct numerical experiments to note that:

- if $\delta_p \leq 0$ then the discriminant is positive and the system is non-hyperbolic;
- modelling δ_p with the models used previously[7] yields negative discriminants, giving a hyperbolic system of equations;
- lower-density liquids give more negative discriminant values and the system requires less interfacial pressure for hyperbolic character;
- higher-pressure systems give more positive discriminant values and the system requires more interfacial pressure for hyperbolic character.

(625g) Transient CFD Simulation of Two-Phase Liquid-Liquid Flow in an Oscillatory Baffled Reactor

Jing Ma, Cheng Liu, Cunting Fu, Xiuqin Dong and **Minhua Zhang**, R & D Center for Petrochemical Technology, Tianjin University, Tianjin, China

Oscillatory baffled reactor is a new type of chemical process intensification equipment. Periodic vortex is generated in the reactor due to its constricted baffle geometry and periodic oscillatory condition, which provide both axial and radial mixing in the reactor, increase the interface area of immiscible two-phase, improve the residence time distribution of dispersed phased, and control particle size and its distribution.

In the present work, a combination of computational fluid dynamics(CFD) and population balance model(PBM) was applied to study the effect of oscillatory condition (the oscillation amplitude and the frequency), baffle orifice ratio and baffle spacing on the flow field and mass transfer characteristics in an oscillatory baffled reactor.

The results showed that, the average velocity increased linearly with the increase of oscillatory Reynolds number in one oscillatory cycle, the oscillatory condition had little effect on the volume distribution of dispersed phase. Both the drop size and size distribution greatly decreased with the increase of oscillatory Reynolds number. A correlation between sauter mean diameter, , and the energy dissipation was established as . The results also indicated that the oscillation amplititude had more influence on the flow filed and droplet size distribution than the oscillatory frequency.

The effect of baffle orifice of 0.21, 0.25, 0.31 and baffle spacing of 1.0D (D, the reactor diameter), 1.5D, 2.0D on the flow field and drop size distribution was investigated respectively. The optimal baffle orifice was 0.25 and the optimal baffle spacing was 1.5D. Not only the information of the effect of oscillatory condition and geometry on the two-phase flow filed and mass transfer characteristics were obtained by the simulation, but also the results provide a theoretical basis for reactor design and geometry structure improvement.

AUTHOR INDEX

AUTHOR INDEX